ELECTRICAL WIRING AND REPAIR

A Guide to Improving and Maintaining Residential Electrical Systems

Martin Clifford

PRENTICE HALL, Englewood Cliffs, New Jersey 07632

Library of Congress Cataloging-in-Publication Data

Clifford, Martin
    Electrical wiring and repair : a guide to improving and
maintaining residential electrical systems / Martin Clifford.
        p.   cm.
    ISBN 0-13-247867-6
    1. Electric wiring, Interior.   I. Title.
TK3285.C58   1991
621.319'24--dc20   90-33840
                                                         CIP

Editorial/production supervision
 and interior design: *Carol Atkins*
Cover design: *George Cornell*
Manufacturing buyer: *Margaret Rizzi*

 © 1991 by Prentice-Hall, Inc.
A Division of Simon & Schuster
Englewood Cliffs, New Jersey 07632

The publisher offers discounts on this book when ordered
in bulk quantities. For more information, write:
    Special Sales/College Marketing
    Prentice-Hall, Inc.
    College Technical and Reference Division
    Englewood Cliffs, NJ 07632

Note: The material in this book may be subject
      to local codes and regulations. Check with
      local building officials when planning your
      project.

Printed in the United States of America

10 9 8 7 6 5 4 3 2 1

ISBN   0-13-247867-6

Prentice-Hall International (UK) Limited, *London*
Prentice-Hall of Australia Pty. Limited, *Sydney*
Prentice-Hall Canada Inc., *Toronto*
Prentice-Hall Hispanoamericana, S.A., *Mexico*
Prentice-Hall of India Private Limited, *New Delhi*
Prentice-Hall of Japan, Inc., *Tokyo*
Simon & Schuster Asia Pte. Ltd., *Singapore*
Editora Prentice-Hall do Brasil, Ltda., *Rio de Janeiro*

To the *new* Free Public Library
of Lakewood, New Jersey,
and its staff and supporters.

Contents

2   ELECTRICAL JOINTS, CABLES, AND SOLDERING   30

3   ELECTRICAL COMPONENTS   73

4   THE "HOW TO" OF HOME WIRING   130

Preface

There is no home that can be said to be adequately wired. If the home is an older type, there will be an increasing need for more branch electrical circuits. Heavier wire gauges may need to be substituted for existing wires, more convenience receptacles (outlets) may need to be installed, more electrical switches may be in demand, or older switches may need to be replaced. Part of or possibly the entire lighting system may need updating.

The homeowner is thus presented with a choice—either make use of the services of a professional electrician or become a do-it-yourself electrician. And that leads directly to the purpose of this book. Many of the electrical changes that need to be made can be done by the homeowner. Even in those instances in which professional services are required, a knowledge of electrical changes and installations may help keep professional service charges at a lower level than otherwise.

This book performs another function. Since it supplies the information needed for in-home and outside-the-home electrical wiring, it serves as a text for those who want to become professionals. It covers a wide gamut of electrical projects and is enough to get the novice electrician started on his career.

Mathematics can be associated with electrical wiring, installation, and repair. It can also be omitted, and that is the path this book takes. There is no math. The emphasis is on two concepts: "What is the problem?" and "What are the practical steps to take?"

Electrical work deals with three basic electrical concepts: voltage, current, and resistance. These must be considered by every electrician and homeowner and are covered in Chapter 1. Voltage, current, and resistance are related and are always found together. While this chapter can be considered (and is) theoretical, it is needed to form a basis for understanding wiring and electrical repair. Electrical work requires a supply of tools, many of which are used in the home for other purposes. Chapter 1 also discusses tools, since using the correct tool will often make wiring changes either easier or possible.

An electrical system consists of power supplied by a utility and appliances to be operated by that power. The connecting link consists of wires that form cables. Those wires must be cut to fit—that is, the length of the wire must be cut to its correct size. Further, connections may need to be made along the length of that wire, so joints must be used. There are a number of different types, and a choice is supplied in Chapter 2. Also included in Chapter 2 is detailed information on soldering. While soldering can often be omitted, in some installations it helps make better connections. This chapter also includes detailed information on the kinds of wires that may be used, essential since wiring is usually specified by municipalities.

Every homeowner and professional electrician is always faced with the need for making decisions about the electrical parts to be used. The number of parts is staggering, and even well-stocked electrical parts houses may not have all of them available. Chapter 3 supplies a broad-range description of those most commonly used. It will be helpful to read this chapter and come to some decision prior to shopping. Using the right components will help expedite the electrical job waiting to be done.

Chapter 4 can be regarded as the "nuts and bolts" of this book, for it explains in detail the scores of electrical jobs to be done in the home. Some of these offer the home owner or electrician a choice of the changes to be made. They can also be modified to meet individual requirements.

Chapter 4 is the "how to" of electrical wiring, and it has a three-fold purpose. It can be used for maintenance of the home electrical system, for making repairs, and also for taking care of present and anticipated electrical needs. Keeping the home electrical system up-to-date is an on-going process and, once the technique of making electrical changes is understood, is basically easy.

Chapter 5 discusses the types of lighting systems that can be used and contrasts incandescent versus fluorescent lighting, explaining problems encountered with each. Both types of light systems have their advantages and disadvantages. This chapter also supplies troubleshooting information.

Outdoor wiring, described in Chapter 6, supplies information about a topic that is assuming ever greater importance. Outdoor lighting is helpful in the prevention of accidents that can occur outside a darkened home, and adds to home security. This is one area in which electrical changes for the home are becoming more common. Further, outdoor lighting contributes substantially to the value of a home and can be used to highlight the beauty of outdoor landscaping.

Chapter 7 describes the types of motors commonly found in the home. The number of motors has grown extensively in recent years, and they are used in a variety of appliances: vacuum cleaners, compact disc players, turntables, washing machines, antenna rotators, pencil sharpeners, typewriters, computer systems, and paint sprayers—this list is by no means complete. Motors require maintenance, and this is often quite simple. This chapter details symptoms and the steps to take to keep motors in good running condition.

No electrical component or wiring lasts forever. Parts wear out or become out of style, become ever more noisy or operate intermittently. Chapter 8, with its emphasis on maintenance, calls attention to the need for making electrical repairs.

Proper maintenance can help prolong the life of an electrical system and in so doing helps make the home more electrically convenient. Maintenance of the system contributes materially to electrical safety and adds to the value of the home.

ACKNOWLEDGEMENTS

I would like to acknowledge with thanks and appreciation the contribution of electrical data, technical information and drawings supplied in the literature and catalogs sent to me by the companies listed below.

      Brooks Electronics, Inc.
      Edison Lighting
      Florida Power & Light
      Gilbert Manufacturing Co., Inc.
      Hubbell Incorporated, Wiring Device Division
      Jersey Central Power & Light
      Leviton Manufacturing Co., Inc.
      Mulberry Products, Inc.
      Sears Roebuck & Co.
      Weller Co.
      The Wiremold Company

Martin Clifford
North Lauderdale, Florida

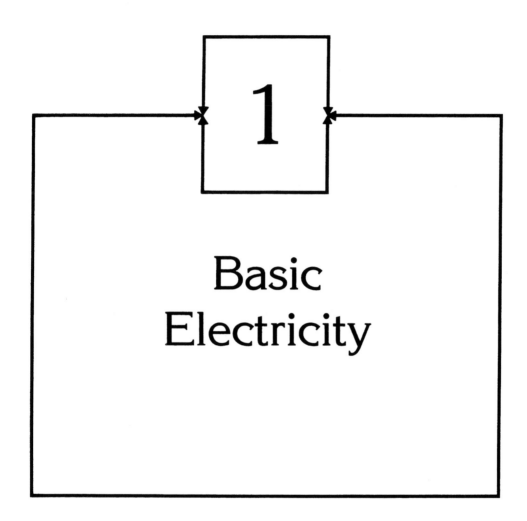

1

Basic
Electricity

Because of the large number of appliances used in the home, it has become essential for its occupants—and that means men, women, and adolescent children—to become acquainted with those appliances from an electrical point of view. It means knowing the kind of wire to use, learning how to connect wires, and it often means acquiring the know-how for updating the wiring system in the home. It also means learning the language of electricity, and determining how to use a few simple tools for electrical work. Some of these jobs are simple, others are less so. Working with electricity is often easy but can be frustrating. Electrical knowledge can also be both a money and a time saver.

WHEN YOU NEED PROFESSIONAL HELP

There are times when it becomes necessary to call for professional help—when in-home electrical repairs require expertise and the use of tools not ordinarily found in the home. However, even here a knowledge of electricity permits better communication and understanding between the electrician and the nonprofessional. But it does more than that, for it is just as important to know your electrical limitations and when to call for the services of a licensed electrician. In some instances, outside help may be necessary to meet the electrical standards and laws of your community.

IMPROVING THE ELECTRICAL VALUE OF YOUR HOME

Electrical maintenance and improvements raise the value of a home, and many of the improvements do not require the assistance of a licensed electrician. Some of these are nothing more than adding another outlet, changing an outlet from a simplex to a duplex type, putting a new plug on a line cord, or augmenting existing wiring or adding new wiring. As an added bonus, the successful completion of an in-home electrical project results in more knowledge and confidence, possibly leading to the ability to cope with more complex changes.

VOLTAGE

There are three basic words used in electricity, and these are voltage, current and resistance. *Voltage* is electrical pressure, and while the word "voltage" is common, it is also known as *electromotive force*, abbreviated as EMF, or as *electrical potential* (or simply, potential), or as *potential difference* (PD).

Although the word "voltage" is widely used, it is also widely misunderstood. The term is also commonly accompanied by various myths incorporated in statements such as "This line carries 115 volts," implying that the voltage is gliding smoothly and effortlessly along a wire. Voltage is simply electrical pressure and is comparable mechanically to the pressure you put on a wall when you lean against it. Another myth is that voltage cannot exist without an accompanying current. A battery supplies an electrical voltage, but you can carry that voltage in a pocket without current flow. An AC receptacle in the home has an electrical pressure, but no current flows from it until some electrical device is plugged in.

Units of Voltage

The basic unit of voltage is the *volt* and is the reference against which all other amounts of voltage are measured. A *millivolt* is a submultiple of a volt and is equal to one one-thousandth of a volt. It takes one thousand millivolts to obtain the equivalent of a volt. Another submultiple is the *microvolt,* or a millionth of a volt. It takes a million microvolts to constitute a single volt.

Moving in the other direction, one of the multiples is the *kilovolt,* equivalent to a thousand volts. Still another multiple is the *megavolt,* or million volts.

For in-home electrical work the most common unit is the volt, although submultiples and multiples may sometimes occur. Volts, millivolts, microvolts, kilovolts and megavolts are part of any electrician's vocabulary, whether that electrician has amateur or professional status.

Source Voltage

Electrical devices of all kinds require an operating voltage, often referred to as a source voltage. That voltage can be obtained from a battery or from an AC power outlet. It is identified as a source voltage to distinguish it from various voltages used inside the electrical device. The source voltage delivered to the device is often divided into smaller amounts inside the device, each of which may also have its own name.

CURRENT

Eiectrical pressure or voltage can result in the movement of extremely tiny particles called *electrons.* The sum movement of these electrons is called a current of electricity or sometimes simply a *current.*

The Load

Any device that requires current and is connected to a voltage source, whether that source is a battery or supplied by a receptacle, is referred to as a *load.* Loads can be characterized as light or heavy. A *light load* is one that requires a small amount of current; one that makes a strong current demand is *heavy.* These terms are relative. A 10-ampere washing machine motor could be considered a heavy load; a small fan a light load. Voltage sources do not supply current until they are loaded.

The basic unit of current is the *ampere,* and this is the reference against which current multiples are measured. A *milliampere,* a submultiple, is a thousandth of an ampere. It takes a thousand milliamperes to form an ampere. Another current submultiple is the *microampere,* or millionth of an ampere. It requires a million microamperes to constitute a single ampere.

Unlike voltage units such as the kilovolt and megavolt, these multiples are not applied to in-home electricity. Terms such as the kiloampere and mega-ampere are not used.

Types of Current

While an electric current can exist in a number of different forms, there are two basic types: direct current and alternating current, abbreviated as DC and AC. *Direct current* is supplied by a battery or by an electronic power supply. *Alternating current* can be taken from power receptacles in the home. Both DC and AC are widely used, are essential, and have different characteristics, but one cannot be said to be superior to the other.

Ampacity

This can be defined as the amount of current a conductor can carry continuously without exceeding its temperature rating. *Ampacity* is a function of cable size, insulation type and the conditions of use.

Polarity

Figure 1-1 is a fundamental DC circuit, consisting of a battery used as a voltage source and a lamp used as a load. The direction of current flow is represented by arrows. These show that the electric current flows from the minus terminal of the battery, through the load, returning to the plus terminal. *Minus* and *plus,* also referred to as *negative* and *positive,* are represented by symbols such as (−) for minus and (+) for plus. The terms minus and plus, or their symbols, represent the *polarity* of the voltage source.

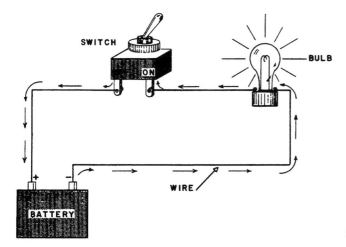

Figure 1-1   Basic DC circuit.

The flow of current does not stop at the terminals of the voltage source. When the current reaches the plus terminal, it continues through the battery, reaches the minus terminal, and exits once again in its movement toward and through the load. While the current is said to flow from minus to plus, the movement of current is comparable to that of a wheel. All parts of the wheel get into motion simultaneously. The minus-to-plus concept simply indicates direction. If the current always flows in the same direction, it is said to be DC, even if the amount of current should fluctuate because of changes in the load.

The direction of current flow can be changed by transposing the connections to the terminals of the battery. The result would still be DC, and the current would still flow from the minus terminal, through the load, to the positive terminal.

Alternating Voltage and Current

Figure 1-2 consists of a battery, a load, and a polarity reversing switch. When the switch is in one position, current will move from left to right through the load, the lamp, and when in the other position, will flow from right to left. If the switch is

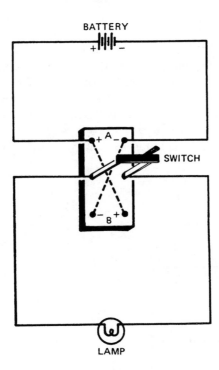

BATTERY

SWITCH

LAMP

Figure 1-2 Use of double-pole, double-throw (DPDT) switch can change current flow through the lamp from DC to AC.

operated on a continuous basis, the current will move back and forth; that is, its direction will alternate. It could then be referred to as an alternating current, but with that current still supplied by a battery.

This behavior can be represented by a graph (Figure 1-3) showing voltage versus time. When the switch is first closed, the full voltage of the battery is applied to the load, as shown by vertical line A-B. As long as the switch remains closed, the voltage remains constant, B to C. When the switch is opened, the voltage on the load drops to zero, C to D. If the switch is then closed, but in the opposite direction, the voltage rises from D to E. As long as the switch remains closed, the voltage will remain constant, E to F. If the switch is then opened, the voltage across the load will drop to zero, as indicated by the vertical line from F to G. The fact that a voltage exists above and below the zero-voltage line shows a reversal of voltage polarity.

While the graph indicates the reversal of voltage polarity by operating a switch, the same comments could be made about the flow of current, and the same graph could be used. Because of the switch the voltage applied to the load is alternated and so is the current. The voltage (and its polarity) at the terminals of the battery remains unchanged.

The battery-switching circuit demonstrates that it is possible to get an alternating voltage and current (AC) from a DC source, such as a battery. Except as a learning device, the circuit is impractical and there are easier ways of obtaining AC.

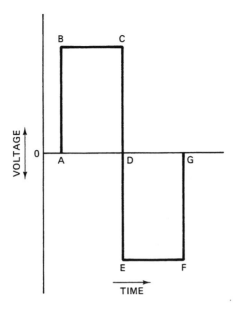

Figure 1-3 Type of current produced by a battery and a voltage-reversing switch.

AC is supplied to homes through the use of a generator, a machine whose output voltage changes its polarity regularly. The graph in Figure 1-4 shows how the voltage changes.

Instead of rising abruptly, the voltage, starting at zero, gradually rises to a peak or maximum. It then decreases automatically following a gentle slope until it becomes zero. At this time the polarity changes and the voltage increases to a peak, following which it slopes down to zero. This complete series of events is called a *cycle*, and in a home there are 60 such cycles per second, every second. A complete

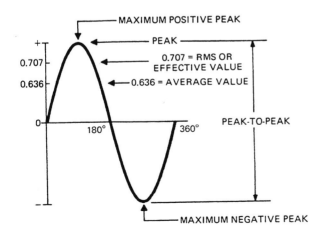

Figure 1-4 Characteristics of a single-phase sinusoidal waveform. The zero reference line is measured in degrees although time, usually in seconds, is sometimes used.

cycle is also called a *hertz,* abbreviated as hz. The number of cycles per second is referred to as *frequency.* In-home electric appliances may have a name plate indicating that the device is to be used with AC having a frequency of 60 hz. Other power line frequencies are also used, notably 50 hz, but are much less common.

The AC Waveform

It is easy enough to measure a DC voltage, for it remains constant. AC, though, changes every moment, presenting a more difficult problem. There are two points, though, at which the voltage does remain momentarily fixed, and these are the positive and negative peaks. Since the negative peak has the same value as the positive peak, the positive peak is chosen as a reference. That peak is 171 volts, and twice during each cycle the voltage available at receptacles reaches this amount.

LINE VOLTAGE

The voltage available at an outlet or receptacle is variously listed as 110, 115, 117, 121, or 125. There is no standardization, nor is it likely or possible. An appliance carrying a label indicating 125 volts AC will work if the voltage is 110. An electric light bulb marked for 110 volts can be used for a line voltage of 125.

The line voltage is 70.7 percent of the peak voltage supplied by the AC generator of a public utility. That peak voltage can vary, depending on the regulation of the generator (its ability to maintain a constant voltage level under varying load conditions), the extent of the load, the distance of the receptacle from the utility, and the ability of the entire electrical system to tolerate overload conditions. If the load is extremely heavy, as, for example, the simultaneous use of air conditioning systems, the line voltage could drop below 110. A condition of *brownout* occurs when the voltage is so low that electric lights barely work. A *blackout* exists when the voltage is so low, or possibly zero, that lights and appliances do not work at all.

Line Voltage Purity

At one time there was no concern about line voltage purity, for the subject didn't exist. Electricity was used for home lighting, percolators, toasters, broilers, washing machines, and dryers and these were not affected by transient line voltage changes. But in-home electrical equipment has changed. Television sets, touch lamps, VCRs, for example, and more particularly computers, now use integrated circuitry and other solid-state components. These appliances are very sensitive to changes in line voltage changes (and can be damaged by them), even those lasting for a fraction of a second. Line voltage in the form of momentary spikes can be produced externally, possibly by thunderstorms, or internally by the feedback of voltage into power lines by washing machines, refrigerators, and oil burners when these cycle on and off.

Devices such as television sets and computers can be made free of line-voltage damage by using surge suppressors. These do not conserve electrical energy nor do they offer any relief from brownouts or blackouts. There are various kinds of surge suppressors. One of the least expensive can be plugged into a wall receptacle. The equipment to be surge protected is then plugged into the surge suppressor. Some surge suppressors are built into strip receptacles and so can protect a number of devices.

As evidence of line voltage surges touch lamps may turn themselves on; digital clocks on VCRs or microwave ovens will sometimes stop or flash. Another safeguard is to have your computer put on a dedicated circuit. This is a branch line with no receptacles other than the one used for the computer.

ELECTRICAL POWER

A bill for the use of electricity isn't based on voltage or current alone, but on a combination of these two, plus a third factor, time. The combination of voltage and current is known as *power.*

The basic unit of electrical power is the *watt.* Submultiples are the *milliwatt,* or thousandth of a watt, and the *microwatt,* or millionth of a watt. A more practical unit for in-home power use is the *kilowatt* (kw) or thousand watts. Another multiple is the *megawatt,* or million watts.

CURRENT AND WATTAGE

There are two levels of voltage used in the home: 120 and 240 volts. As far as current is concerned, it can range from fractions of an ampere to multiamperes.

While an electric bill is based on the amount of wattage used multiplied by the length of time, the important factors are current and time. Electric power in watts is equal to voltage multiplied by current, but the voltage is constant.

For an input of 120 volts and a current of 1 ampere, the power is $120 \times 1 = 120$ watts. If the current is 2 amperes, the power increases to $120 \times 2 = 240$ watts. And for 3 amperes the power usage rises to $120 \times 3 = 360$ watts.

These numbers, though, must be multiplied by time. If we assume 2 hours for each of these examples, the first would be 120 watts $\times$ 2 = 240 watt hours; the second would be 480 watt hours, and the third would be 720 watt hours. These two variables, current and time, give the consumer some control over electric bills. The consumer can keep time usage as low as possible and avoid, if possible, current-hungry electrical appliances. Even electric light bulbs if left on unnecessarily and for long periods of time, especially multiple, higher wattage types, contribute substantially to the final amount of an electric bill.

The Advantages of Higher Voltages

There are two voltage levels used in the home. The most common, of course, is a power line voltage of 120. However, 240 volts is also used for current-hungry devices. An appliance connected to the 120-volt line and requiring a current of 10 amperes will have a power demand of $120 \times 10 = 1200$ watts. If the line voltage is doubled, increased from 120 to 240, the current requirement is cut in half, $240 \times 5 = 1200$ watts. The power usage remains the same. The advantages of using the higher voltage and lower current means that a thinner wire, one having a higher wire gauge, can be used. Also, there will be a smaller power loss in that line.

CURRENT RATINGS OF TOOLS

Homes today are electric, and in some instances they are designated as "all electric," generally signifying that even the heating power plant is electric, indicating that no oil, gas, coal, or other type of fuel is used.

This move toward complete electrification also applies to tools of all kinds: electric drills, belt sanders, grinders, jig, saber and circular saws, and so on.

Data about electric tools, including separate motors, are supplied on the data plate mounted on the tool body, as indicated in Figure 1-5. There is no data standardization, but it may include tool identification, model number, the input voltage, the type of input voltage (whether AC or DC), the power line frequency, the motor

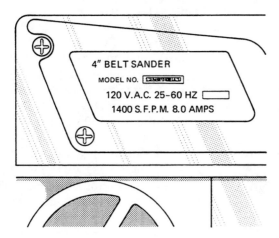

Figure 1-5 Data plate on an electrical power tool.

speed, and the current requirement. The frequency may be a range, such as 25–60 hz, although it is usually 60 hz.

For electrical power tools in the home the current rating is ordinarily 2 to 13 amperes. Usually larger tools have a greater current demand, but even if two tools seem to be about the same, the one having the higher current rating will be more powerful.

Power Precautions When Using Electrical Tools

It is easy to overload a branch power line with an electrical tool. While its motor may be rated as having a current demand of only 4 amperes, the starting current of its motor may be as much as five times this amount. It will be helpful to use a slow-blow fuse for such applications.

If the electrical tool has a rating of just a few amperes, it can be connected to any branch, keeping in mind that the branch should not be overloaded with the simultaneous operation of a number of small appliances. Use a branch circuit intended for 15-ampere operation.

An electrical tool requiring up to as much as 10 amperes can also be operated on household circuits, provided these are not, at the same time, supplying power to other appliances such as freezers and air conditioners. Use a branch circuit intended for 20-ampere operation.

Heavy duty tools, those requiring current in the order of 12 or 13 amperes, can be operated from the 20-ampere line, which pushes this branch line close to the limits of its capability. It is better to use the 30-ampere branch, but even here some caution is required about using other current-hungry appliances on the same line at the same time.

Electrical Tool Suitability

Electric tools intended for in-home use may be described as "domestic" or "utility." Those for commercial use may be labeled "industrial," "commercial," or "builder." These are generally intended for manufacturing use and may have current demands that cannot be easily met by branch power lines in the home.

Power Tool Motors

Power tools for in-home use have either a rotary or a vibrating-type electric motor.

The rotary motor has a shaft that revolves through 360 degrees repeatedly. It is often internally connected to a series of gears for controlling speed and output torque. Such motors are often brush equipped. Eventually these brushes wear out,

causing excessive sparking and possible motor failure. The brushes can be replaced
but require a duplicate, obtainable from the manufacturer or his dealers.

A vibratory motor, such as those used in jig saws and finishing sanders, is so
called since it supplies a back and forth motion. This tool uses a motor that is not
equipped with brushes and is intended for light duty. Its current requirements are
much less than those of heavy-duty power tools.

ELECTRICAL HOUSEHOLD APPLIANCES

Because in-home electrical power is so readily available, so convenient, and such a
labor saving plus, the number of home-use appliances is astonishingly large. All of
them are susceptible to abuse, overwork and lack of maintenance.

The list in Table 1-1 is not complete, and it is doubtful if it could ever be, but
at least it supplies some indication of the extent to which households are electrically
oriented and can include: individual room air conditioners, one- and two-zone overall air conditioning, blenders, coffee makers, dishwashers, clothing dryers, hair dryers, garbage disposal units, irons, waffle irons, can openers, ovens, toasters, corn
poppers, food processors, refrigerators, stoves, hot plates, vacuum cleaners, paint
sprayers, power tools, fans, pencil sharpeners, lights, clocks, room ventilators, de-
humidifiers, furnaces, sandwich grilles, shoe polishers, ice cream makers, frying
pans, trash compactors, portable heaters, sewing machines, woks, Christmas tree
lights, security systems, doorbell and chime systems, plate warmers, broilers, food
freezers, and water heaters.

TABLE 1-1 APPROXIMATE POWER REQUIREMENTS
OF HOME APPLIANCES

Appliance	Average power in watts for 120-volt line
Air conditioner, room type	1000
Blanket, electric	250
Blender	300
Broiler	1500
Can opener	100
Central air conditioner (240 V)	5000
Clock, electric	2 to 3
Coffee maker, percolator	1000
Dehumidifier	500
Dishwasher	600 to 1000
Drill, electric	300
Dryer, clothes	4000
Fan, 8-inch	30
Fan, 10-inch	35
Fan, 12-inch	50
Fan, attic type	400

TABLE 1-1   (continued)

Appliance	Average power in watts for 120-volt line
Fluorescent lights	15 to 40
Food mixer	200
Freezer	500
Fryer	1200
Furnace, oil	900
Garbage disposal unit	500
Hair dryer	300
Heater, radiant	1200
Heating pad	50 to 75
Hot plate	1000
Hot water heater (240 V)	2500
Humidifier	500
Infrared heating lamp	350
Iron, clothing	1500
Iron, soldering	30 to 250
Lamps, incandescent	4 to 200
Microwave oven	650
Mixer, food	200
Motors	20 to 350
Radio receiver	10 to 150
Range, all burners and oven turned on	8000 to 16,000
Razor, electric	8 to 12
Refrigerator	300
Roaster, electric	1200
Rotisserie	1300
Saw, radial	1500
Sewing machine	60 to 90
Stereo hi-fi system	300 to 500
Sun lamp	300 to 500
Television receiver	100 to 500
Toaster	1200
Vacuum cleaner	950
Waffle iron	600 to 1000
Washing machine, clothing	1300
Water heater	4500

POWER COST

Power cost is based on kilowatts multiplied by the total number of hours the power is used and is referred to as the *kilowatt-hour,* or kw hr. In-home appliances may carry a name plate indicating the number of kilowatts, or fractions of a kilowatt, required.

The greatest contributor to large electric bills are those appliances that require relatively large amounts of current. Such appliances would include electric heaters, clothing irons, heavy-duty soldering irons, toasters and water heaters. However, time is also a factor, and a half-dozen electric light bulbs, even if rated at only 100 watts, can represent a substantial cost if they are kept on for long periods of time.

The Kilowatt-Hour Meter

Electric power bills are based on data supplied by a kilowatt-hour meter located inside the home or outside. The meter makes two measurements simultaneously: the number of kilowatts used and the elapsed time in hours.

There are two types of kilowatt-hour meters: the cyclometer and an older type generally referred to as a watt-hour meter. The cyclometer resembles a small box with a rectangular opening through which dial numbers can be seen. The *cyclometer* supplies a direct reading in kilowatt hours.

To read the watt-hour meter, it is necessary to take two readings: one at the beginning of the month, the second at the end of the month. Subtracting the initial reading from the final one supplies the amount of power usage during this time period.

The upper four-dial drawing in Figure 1-6 is the initial reading and represents 6019 kilowatt hours. The reading a month later is 6299 kilowatt hours. The difference between the two readings is 280 kilowatt hours. Multiplying this number by

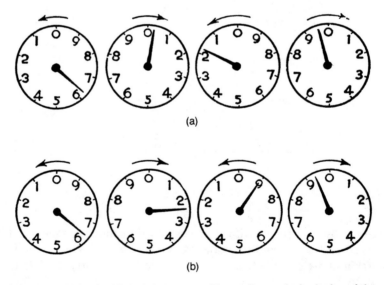

(a)

(b)

Figure 1-6 Dials of a kilowatt-hour meter. The reading at the beginning of the month is subtracted from a reading at the end of the month. (a) Start reading; (b) end reading.

the cost per kilowatt hour supplies the total amount of the electric bill, but the amount does not include any other charges that may have been added.

RESISTANCE

When a current flows through an electrical conductor, such as a wire, it encounters resistance, sometimes referred to as *electrical friction*. The basic unit of resistance is the *ohm*. Submultiples are rarely used. Multiples are the *kilohm* or thousand ohms and the *megohm* or a million ohms.

All conductors used as electric wiring have resistance. The amount of resistance varies directly with the length and inversely with the cross-sectional area. Thus, the longer the wire, the more resistance it presents to the flow of an electric current through it. The thicker the wire, the greater its cross-sectional area and the lower the amount of resistance. The ideal arrangement would seem to be the use of a wire that is as thick as possible and as short as possible, but this is often difficult or impossible.

VOLTAGE DROP

The resistance of a wire has an effect on the actual amount of voltage delivered to an appliance. If the voltage at a receptacle is supposed to be 110 volts and the voltage loss across its connecting wires is 15 volts, the actual voltage supplied to the receptacle is the source voltage minus the line voltage loss, or, in this example, $110 - 15 = 95$ volts. This results in three possibilities. The appliance may work well with 95 volts, it may work poorly, or it may not function at all. There is also a voltage loss in the connecting wires between a receptacle and an appliance. The amount of this voltage loss can be measured by testing the voltage present at the receptacle and also at the input to the appliance. The difference between the two readings, subtracting the smaller number from the larger, indicates the amount of voltage loss.

POWER LOSS

When an electric current flows through a conductor, there cannot only be a voltage loss (also called a voltage drop) across that line, but a power loss as well. The amount of power loss depends on the amount of current flowing through the line and also on the resistance of that line. To calculate the power loss, the amount of current (in amperes) is multiplied by itself and then is also multiplied by the resistance (in ohms). The result is the power loss (in watts).

The power lost in a line is in the form of heat. The most dramatic example of power loss is in an electric light bulb. The purpose of the bulb is to supply light. Light bulbs can get so hot, however, that they can be impossible to touch. That heat is a waste, but there is little or nothing an individual can do about it, other than using a lower wattage bulb or substituting a fluorescent type. Wire is another matter. The power loss can never be completely eliminated, but it can be reduced by using the right size wire and by some prior planning to keep the wire as short as possible.

Power loss exists wherever current is used. If the plug connecting an electrical device to a receptacle gets unreasonably hot or feels unusually warm, you have an indication of a problem somewhere, starting with a poor plug connection but leading to a suspicion that the electrical device or its connecting wire are at fault.

BASIC ELECTRICAL CIRCUITS

There are three basic electrical wiring circuits: series, parallel (also known as shunt), and series-parallel.

The Series Circuit

In this circuit, shown in Figure 1-7, the same amount of current passes through each of the components. The voltage that may be applied is determined in advance by

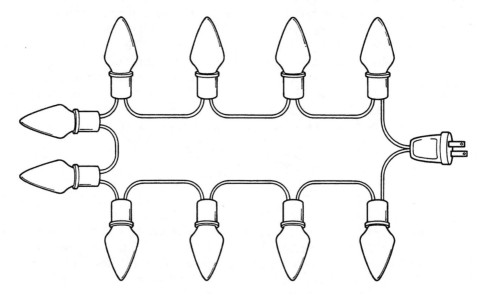

Figure 1-7 Series circuit. The same amount of current flows through each of the lamps.

the manufacturer and is indicated on the product. As an example, if a pair of light bulbs are rated at 120 volts, two of them can be connected in series across a 240-volt line. If the voltage is less than 240, the bulbs may light poorly or not at all; if more than 240, the bulbs will light excessively brilliantly and may burn out.

There are 10 bulbs wired in series in Figure 1-7. If the line voltage is 120, then each of the bulbs should have a 12-volt rating.

The problem with a series connection is that a single open component is equivalent to an open switch, so none of the other components will work. A series connection is commonly used with Christmas tree lights but otherwise has few practical applications.

The Parallel Circuit

In this circuit all of the components to be connected must have identical line-voltage ratings, and for in-home use this is ordinarily 120 volts. Each of the components may have different current requirements, but this is generally indicated as a wattage rating marked somewhere on the component. The higher the wattage rating, the greater will be its current demand. (Figure 1-8)

Unlike series-connected parts, units can be disconnected from a parallel-wired branch without affecting any other component. The total current drain from the power line is the sum of the current demands of the individual units when these are turned on.

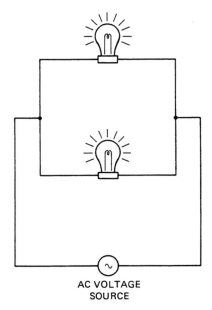

AC VOLTAGE
SOURCE

Figure 1-8 Parallel circuit.

The Series-Parallel Circuit

This circuit is a combination of the series and parallel circuits. Switches are put in series with the power line; receptacles and the components they accommodate are wired in parallel with it.

Pictorial vs Circuit Diagrams

A pictorial diagram shows actual electrical parts as in Figure 1-9a. It is more difficult to draw than a circuit diagram (Figure 1-9b), which relies on electrical symbols (Figure 1-9c). Pictorials are valuable, since they make methods of connections obvious.

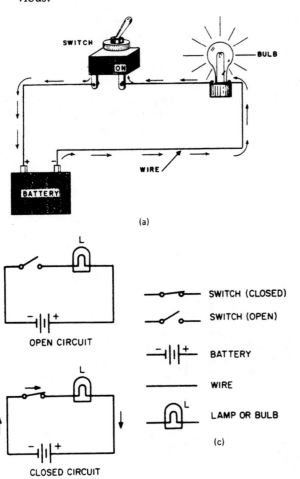

Figure 1-9 (a) Pictorial diagram; (b) equivalent circuit diagram; (c) circuit diagram symbols.

Open Circuit

An open circuit is one through which current does not flow. But be careful! An open circuit does not mean zero voltage. All receptacles used for electrical appliances have voltage present whether those appliances are plugged in or not. A voltage can be present even when it is not supplying current. Open the switch to a lamp or remove the bulb, and an open circuit condition has been created. The voltage delivered to the receptacle into which the lamp is plugged is still there.

Closed Circuit

A closed circuit is the opposite of an open circuit and is one in which current flows to a load. This does not mean that the load, possibly an appliance, is receiving the correct amount of current, nor does it mean the appliance is working properly. All it signifies is that the appliance is connected to a voltage source, that the circuit is complete, that a switch, if any, is closed, and that current is flowing from the voltage source to and through the load. If an appliance works correctly, it is obviously receiving the correct amount of current. If it does not work well, or doesn't work at all, it may be receiving less current than it should.

Short Circuit

A *short circuit* is a type of closed circuit. It means that there is a direct connection across a pair of wires that lead to a voltage source. A short circuit usually demands the maximum amount of current from the voltage source, far in excess of that required for the operation of an appliance. A *leakage current* is a type of short circuit and consists of a path between the wires leading to the voltage source. This path has a much higher resistance than that of a short circuit, so a leakage current is smaller than that of a short circuit. Both short circuit currents and leakage currents are undesirable.

The Conductive Link

The electrical setup in every home consists of three basic parts: (1) a voltage source, (2) various appliances to be operated from that source, and (3) a conductive link between the appliances and the voltage source. The conductive link consists of copper wires between the source and the appliance(s) to be operated.

GROUND

A voltage always exists between two points, and a current flows through two conductors. In an electrical wiring circuit, the two conductors consist of wires color coded black (or red) and white. In an AC circuit, the current moves back and forth through these two while an electrical voltage exists across them.

The white wire, known as the *neutral* wire, may be grounded at one or more points along its length. It can be considered connected in parallel with ground, so the voltage exists between the black or red wires and ground, as well as between the black and red wires and the neutral wire. Touching a water pipe (which is grounded) and a red or black wire will produce a shock just as readily as touching a black or red wire and the neutral wire. An electrical current moving back and forth between the black or red wire, flows not only through the neutral wire but through the ground connection as well.

The problem with using a ground for current conduction is that, unlike the neutral wire, it may be interrupted somewhere along its path. Ordinarily, a water pipe is regarded as a good ground connection, since it is actually located in the earth.

Figure 1-10 shows how this ground connection can be interrupted. The water pipe enters the home but does so through a water meter whose input and output terminals may be electrically isolated from each other. If the ground circuit isn't continuous, the danger of shock is greatly increased. To assure ground continuity, put a metal jumper across the water meter if it isn't so equipped.

The jumper can consist of a number of strands of heavy wire, stranded to supply flexibility. Thoroughly clean the two pipes at the entrance and exit of the water meter, then use a pair of metal clamps to make a good connection to the pipes. The wire joining the two clamps should make a good electrical connection. Since the wire is in the ground circuit, it need not be insulated. This arrangement will make the cold water pipes throughout the home work as an effective ground.

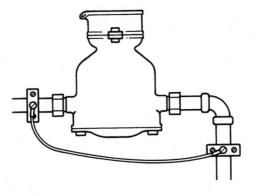

Figure 1-10 If the water meter's input and output are insulated from each other, shunt the connecting pipes with a heavy cable.

GROUNDING APPLIANCES

Some appliances should be grounded, but this precaution doesn't apply to all of them, including toasters, broilers, and electric heaters. These all have one characteristic in common—they have a heating element which is shunted across the branch power line. Touching the heating element when the unit is turned on and also touching a water pipe (ground) or the outer metal shell of the appliance connects the person directly across the power line.

To clean a broiler, a toaster, or an electric heater, be sure to remove its connecting plug from its receptacle. As a further precaution, turn its operating switch to its off position. Make sure the unit is completely cool. Toasters and broilers have a bottom collecting pan. Shaking the units gently will sometimes permit caught food particles to drop into the pan. If not, try to dislodge the food with a small wooden rod or its equivalent. The food collection pan at the bottom of the unit is often hinged and is easily accessible.

THE NATIONAL ELECTRICAL CODE

Yes, it is your home or apartment and your castle, and you can regard it as such, but as far as electrical wiring and the installation of appliances are concerned, there are certain things you may do and others you may not. Electrical wiring is hemmed in by all sorts of restrictions specified not only by your municipality but also by homeowner's insurance policies you buy and pay for.

When you sign an application for a policy, the presumption is that you are aware you are affixing your name to a contract and intend abiding by its terms. If a fire occurs in your home due to some contributory negligence on your part, and if that fire is in some way associated with the improper use of appliances or wiring installation, you may have some difficulty collecting, or if you do collect, you may not get the full amount to which you would otherwise be entitled.

Under the aegis of a number of insurance companies, there is a group known as the National Fire Protection Association. Every few years this group publishes a summary of electrical wiring rules under the heading of *The National Electrical Code.*

The purpose of this code is to provide guidelines for safety in wiring. The reference section of your local library may have a copy, or you can get a copy of the code by writing to the National Fire Protection Association, One Battery March Park, Quincy, MA 02269.

Your existing local municipality code and your state code, if such a code exists, affect home wiring. Also, your homeowner's policy may have a clause that all wiring must be done by a master electrician or must be approved by him.

This does not mean you cannot do your own wiring. It does mean that you, or a professional electrician, must follow certain rules. These rules are not capricious or arbitrary. They are based on substantial experience and are for your own protection, as well as the protection of your neighbors and their property.

Fortunately, you can make numerous electrical repairs; you can replace worn-out electrical equipment; and you can keep your home in tiptop electrical shape by making your own repairs provided you do so in a good way.

UNDERWRITERS' LABORATORIES

Every home electrical repair requires two basic items: labor and materials. There are three ways of handling labor. You can do all the work yourself; you can have a joint working relationship with an electrician; or you can employ an electrician to do all the work. By becoming knowledgeable about in-home electricity you can avoid unnecessary or expensive repairs.

Electrical materials are not only good but are in plentiful supply, and you have a tremendous variety from which to pick and choose. Buying electrical equipment is like shopping in a supermarket. You do best if you know which items represent value.

Many electrical parts carry a label indicating they are approved by the Underwriters' Laboratories (UL). Electrical parts listed in catalogs will frequently show some indication that they are Underwriters' Laboratories approved. You will see the logo or identification symbol of the Underwriters' Laboratories. Use only those parts that are UL approved. UL labels come in a variety of sizes, styles and shapes (Figure 1-11) but if you read the label you will have no doubt about the intent.

UL approval is not a guarantee of top quality but simply indicates that certain minimum standards have been met. You can buy two switches, both UL approved, but with one costing more and of superior quality and workmanship than the other. Both will be safe to use, will work as you expect, but one may last longer, be quieter, and be easier to mount, install, and handle, or may have other features over and beyond the normal functions of a switch. But the less expensive switch will also work and will do so safely.

The Underwriters' Laboratories was set up as a laboratory for making exhaustive tests of electrical products submitted to it by manufacturers. If the components meet all tests, it is then eligible for listing in the "List of Inspected Electrical Appliances" and may then carry a label indicating that such components are UL approved. Manufacturers of electrical products are interested in UL approval, since this gives them a competitive edge over products that have not been approved. This

Figure 1-11 Various UL labels for electrical equipment.

does not mean that an electrical product has UL approval in perpetuity. Products that go through materials or design changes must be submitted for reapproval.

The UL stamp or label means the product you buy has met tests imposed by an outside agency and is for your protection. The UL label may actually be that: a label adhering to the surface of the product. Or, it may be stamped or printed on or be a tiny metallic disc. But whatever its shape, size, color and no matter how it is applied, look for the words or abbreviation, Underwriters' Laboratories or UL.

Following the Rules

The UL tag of approval does not supply carte blanche to use electrical equipment beyond its specifications. If you buy a length of wire rated for two amperes, you may not arbitrarily decide to use the wire with a four-ampere current. UL approval is for a product to be used within the specification limits indicated by the manufacturer. These are the specifications submitted to the Underwriters' Laboratories, together with the material to be tested. One of the functions of the Underwriters' Laboratories is to make sure that electrical products do meet the specifications claimed by the manufacturer.

How Can You Be Sure?

You can always find a small, marginal merchant willing to sell electrical equipment at bargain prices. Often the equipment will be sold with the statement that it is as good as UL approved, or there may be an implication that UL-approved manufacturers have gathered together to stifle competition. That is nonsense. Buy only from a reputable electrical-parts merchandiser. If you buy appliances, make sure you get a warranty, but before you leave the store, make certain you understand for what labor, parts, and length of time the warranty is valid.

When To Make Repairs

How do you know if your home needs electrical changes? If your home is more than 20 years old, it is probably inadequately wired. Two decades ago fewer electrical appliances were used and those that were available didn't have the current demands of today's components. Further, more recent electrical appliances are current-hungry types: self-cleaning ovens, microwave ovens, rug (washing) cleaners, and so on. Many homes now use more than one television receiver, and have track lighting, four-slice instead of two-slice toasters, clothing dryers, and so on. Since the trend is toward more, not fewer, appliances, wiring changes must be made not just to meet immediate needs, but with the heavier current loads of the future in mind.

Is Your Home Wiring Adequate?

How can you decide if your home wiring is adequate? If all the appliances in your home work well and continue to do so even if most or all of them are turned on at one time, if your lighting remains steady and brilliant, if your television picture fills the entire screen and has adequate brightness, if you do not need to replace fuses or reset circuit breakers, then you have a number of indications that your home wiring is adequate.

Don't look on this as a permanent condition. All it means is that you have not as yet reached the limit of the current capacity of your wiring system. All it signifies is that with your existing appliance setup your electrical system is working well. It does not mean you can continuously and arbitrarily keep adding one appliance after the other without any expectation of trouble.

Trouble Symptoms

Here are some trouble symptoms that will supply some clues about the adequacy of your home wiring.

You install a new window-type air conditioner and its performance is disappointing. Further, you have circuit breaker troubles, something that has never happened before.

Your master fuse box has circuit breakers, and you find it necessary to reset these breakers fairly often. If, instead of circuit breakers, your box uses replaceable fuses, you find you must put in new fuses more often. Also, you may be using fuses having higher current ratings than the originals. You are aware that this is the wrong thing to do, but you want to get back to the time when you weren't faced with these problems.

The lights in some of your rooms seem dimmer, particularly when you turn on an air conditioner, television set, or electric range. You replace the electric light bulbs with some having a higher wattage rating but the situation not only does not improve but seems to get worse. At times the lights seem to flicker. This convinces you that the electric light fixture needs to be replaced. Conversely, the lights in one or more rooms seem much brighter when the appliances are turned off. This leads you to the conclusion that electric light bulbs aren't made as good as they once were.

You need to "jiggle" one or more of the switches to get the lights to turn on. This convinces you that the switch (or switches) need to be replaced. After doing so, you are aware that the problem has remained.

You have started to use cube taps at various outlets to accommodate the increasing number of appliances you now have. Now you find that even using one or two cube taps at different outlets is still not enough. You have begun plugging a cube tap into a cube tap. Aside from being very unsightly, you have become aware that this is an indication that you may be overloading a branch power line leading to your circuit breaker box.

Upon removing a plug from an outlet, you touch its prongs and are surprised to find they are warm. Or you touch the insulated wire leading to the appliance and find it is quite hot. This puzzles you and you are convinced that the appliance is defective, and, under the terms of your warranty, you demand and get a replacement. However, with the new appliance, the plug and its connecting cord still get hot.

Your television picture no longer seems as bright as it once was. Your local TV repairman talks about installing a new picture tube even though the one you have is just a few months old and your television set doesn't get much usage. You notice that the picture improves considerably when you turn off a number of appliances connected to the same, or adjacent outlets.

The automatic feature on some of your appliances functions erratically. Your toaster doesn't pop up when it should; your washing machine is now in the habit of skipping a cycle or two.

These are just some of the possible symptoms. If you have a few of these, or similar problems, then at least part of your home wiring system is due for an electrical update.

This does not mean your home must be completely rewired. Possibly just one or two wiring branches from your fuse or circuit breaker box need attention. Even if your home is brand new, the need for updating starts from the moment you move in. The faults may be due to inadequate wiring or appliances that need repairs or replacement.

TOOLS

The number of tools you will need for electrical repairs will depend on whether you have an apartment or your own home. The amount of money you save by doing your own electrical jobs will soon pay for the investment in tools. And most tools can be used for more than just electrical repairs.

Tools for electrical work include screwdrivers, both flat blade and Phillips types, a hammer, long-nose pliers, gas pliers, a soldering iron or gun, diagonal cutters, wire strippers, an electric drill and a set of bits, a hacksaw, and possibly some other tools as well. Actually, no such list is ever complete. This doesn't mean you must buy all the tools mentioned here; get them as you need them.

Fish Wire and Drop Chains

It may sometimes be necessary to pull wires through conduits or behind walls, and for this purpose you can use fish wire. Getting wire through conduit sometimes requires quite a bit of force, so a fish-wire grip is needed.

Fish wire is made of tempered spring steel about ¼ inch in diameter, and you can get it in various lengths. While it is fairly stiff you can push or pull it around bends or elbows of conduit.

SAFETY

Electricity is quiet, and it is this lack of sound that can lull us into a false sense of security. Make no mistake about it—electricity is and can be dangerous. Electrocution requires only a very small current, a few thousandths of an ampere, and just a few seconds. An ideal servant, it makes a terrible master. If you need to work on an appliance, even if it is nothing more than an electric toothbrush, remove it from its power receptacle. Do not depend for your safety on just turning off the appliance's switch. If you work on a fixture, such as a lamp fastened to a wall or suspended from the ceiling, the fact that the lamp is not turned on is meaningless. Remove the fuse or open the circuit breaker controlling that light. Check the light switch to make sure that the lamp is really turned off.

Electricity works with frightening speed. In the case of an accident involving electrical power, the chances of escaping, of pulling away in time, are practically nil. You cannot possibly move as fast as electricity. Precaution is the only safety measure.

Assuming a fuse in good condition or a circuit breaker in its on position, there is always voltage present at the output terminals of a receptacle. Always check to make sure, using either an electrical tester, a volt-ohm-milliammeter (VOM) or a lamp in good working order. Never work on a power line, a receptacle, an electric box, or electrical wiring unless you have turned the power off. In addition, always be sure to turn electrical switches off.

Electricity and water mix, only too well. When working on house wiring, make sure you aren't standing in a wet spot and that your shoes and socks or stockings are dry. If turning off the power deprives you of the light you need for working, use a flashlight.

Electricity and bathrooms make dangerous combinations. Without exception, keep line-powered radios and television sets out of the bathroom. If you must use an electric heater in a bathroom, turn the heater on until the bathroom is warm. Then turn the heater off and get it out of that room. Then, and only then, take your tub or shower.

The first step to take when using electricity is to think safety. Electricity can also cause fires, so if a lamp cord is dried out and cracked, if a plug is hot when you remove it from its receptacle, you have ample warning signals.

If you have a toaster and a slice of bread gets caught, remove the plug before putting your fingers in the appliance. If you have a clothes washer, keep your hands out of the water. If you must transfer damp clothes from a washer to a dryer, use rubber gloves.

Make sure all appliances are adequately grounded when installed. Ask to be shown the grounding connection when you have any appliance put in your home for the first time. Do not be put off with statements such as "automatically grounded." Demand to be shown exactly what automatically grounded means. It is advisable to install ground fault circuit interruptors, especially in a bathroom or laundry room.

The power receptacles can be dangerous for children whose curiosity prompts them to poke around. Get safety cover plates for outlets if you have small children. Curiosity, though, isn't limited to children. If you see an exposed wire, don't touch it until you have learned more about that wire—where and how it is connected and why it should be exposed.

If you have an outdoor swimming pool, or a hot tub, or a jacuzzi, have it thoroughly checked out by a professional. Don't do it yourself. Keep power-operated television receivers and radio receivers or any other line power-operated appliances away from the pool.

Electricity, alcohol and drugs do not work together. Electricity is absolutely unforgiving and isn't known for supplying a second chance.

Do not work on branch power lines in your home, on electrical devices, or receptacles without an adequate amount of light.

Do not plug in and try to operate electrical devices without reading their instruction manuals. Know what the device is supposed to do and how it works before connecting it to a receptacle.

Never join a black lead or a red power lead directly to a white lead. While this will undoubtedly blow the fuse or open the circuit breaker, the flash can produce a body-damaging reaction.

Electricity is often referred to as a servant, and so it is. But it can become the master in a fraction of a second.

Professional electricians do work *hot,* with the power in an on condition, and they do so to avoid the trip to the fuse or circuit breaker box. They take a calculated risk, something the amateur in-home electrician should not do.

Do not ground any part of your body when working with electricity. Do not hold on to metal conduit, to a metal radiator, or to the metal frame of an appliance while using the other hand to work on an electrical circuit.

When disconnecting an appliance, make sure its line switch is turned off. Pull the plug from the receptacle. Never pull the cord.

Never use outdoor lawn tools when the grass is damp or wet. Grass can be damp from early morning dew, not necessarily rain. Cutting damp grass will damage it, while the moisture presents an electrical hazard.

For electrical tools which are used in bathrooms, kitchens, basements and garages where sinks and water conditions exist, use ground fault circuit interruptors (GFCIs). These units will trip under short-circuit conditions and protect the appliance user. GFCIs are becoming commonplace for building code compliance in new homes and can easily be installed in older homes. GFCIs are a must for swimming pool electrical installations.

When buying lighting fixtures, be sure to learn the maximum wattage rating of the bulbs that may be used.

If you have an appliance that sparks, immediately turn off its switch and remove its plug from the receptacle. Never try to force a plug into a receptacle. Some plugs are polarized while the receptacle may not be.

In the event of an electrical fire, shut off the appliance and, if possible, remove its controlling fuse or turn off its circuit breaker. Use a recommended fire extinguisher or common baking soda. Never, but absolutely never, use water on live wires. This is especially applicable to water known to have a high mineral content (hard water).

Never operate an electrical tool while your hands are wet. Never poke or touch the inside of an appliance or tool with your fingers or a metal object unless you are sure the appliance has been disconnected. Be sure to turn off light switches when changing light bulbs. Don't place line cords where they will be in the path of a vacuum cleaner or will supply a tripping hazard or get excessive wear.

Electrical Safety When the Lights Go Out

Step 1. Be prepared. Keep a small supply of items on hand which will be useful in case of a power outage. Organize an outage kit and keep it in an easy-to-find location and make sure all family members know where it is. An outage kit will contain a list of emergency telephone numbers, a flashlight, candles, matches, bottled water, extra blankets or sleeping bags, canned foods, a can opener, and a battery-operated radio receiver. Check the outage kit at regular intervals, replacing batteries when necessary.

Step 2. If weather conditions are such that an outage is possible, turn the controls of your refrigerator to the coldest position. Food will stay frozen between 36 and 48 hours in a fully-stocked freezer if the door is kept closed. Food in a half-full freezer will keep for 24 hours. During the winter, food can be stored in a cold area outside the home. Use caution. Be careful about using uncooked meat and fish or any foodstuff containing cream.

Step 3. Telephone your local electric power utility. Even if you think others may have made such a call, you will be helping to pinpoint the extent of the outage.

Step 4. Until power is restored, turn off all appliances, particularly major appliances such as air conditioners, the electric range and clothes dryer. A single light left in its ON position will signal when power has been restored.

If you try to cook indoors without using the electric range, remember that the two biggest hazards are fire and carbon monoxide. Carbon monoxide is a colorless, odorless gas, and is toxic. Consequently, avoid burning charcoal. Use your fireplace if you must cook.

If You Are on Life Support Equipment

Don't wait for an emergency. Call your local electrical utility if you depend on electrically-operated equipment, such as a respirator or a kidney dialysis machine. While a power utility cannot always avoid a power loss during an outage, they will try to restore power for life-support equipment users on a first-priority basis. Life support equipment should have a battery backup system for emergency operation.

Emergency Provisions for the Service Box

If the service box uses fuses, keep an assortment on hand of each type used in the box. This isn't necessary if the box uses circuit breakers.

Keep a flashlight mounted near the box. The best type is the one using rechargeable batteries. Set up a schedule for recharging the batteries. Use the flashlight for emergency purposes only and make sure all members of your family know its location.

Electrical Precautions Outside the Home

Keep ladders away from electrical cables, especially if the ladders are made of aluminum.

If you spot a fallen power line, telephone your local power utility. If you cannot reach them, telephone the police or fire departments. This situation is especially dangerous for children. Do not leave the fallen-wire area until the utility repairmen, police, or firemen reach the scene. The minute you see a fallen power line you are involved, whether you want to be or not.

Stay away from the fallen power line and anything it may be touching, such as a fence, puddles of water, or an automobile. Wire or metal structures can be electrified by a fallen wire some distance away.

If you are inside a vehicle which comes in contact with a fallen wire, try to drive clear of it. If that fails and you are not injured, you should remain in the vehicle, as this is the safest place to be under the circumstances. On occasion, evacuation from the vehicle may be necessary due to fire. If so, open the door of the vehicle, turn facing outward without touching the ground or any articles outside the vehicle, stand on the doorsill (rocker panel) and jump free of the vehicle. This action will lessen the chance of your completing an electrical circuit between the vehicle and the ground. Under no circumstances should you step out of the vehicle! The key is to jump or keep clear of the vehicle before touching the ground or anything outside the vehicle.

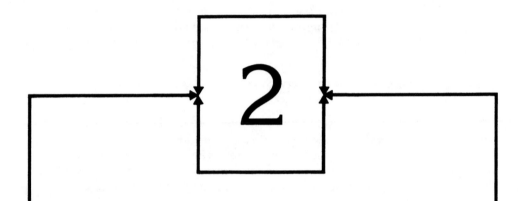

Electrical Joints, Cables, and Soldering

An electrical current, whether weak or strong, always makes a round trip from its voltage source to and through a load and then back to that voltage source. No current is ever lost, and the same amount of current exiting from the source always returns to it. The concept that an electrical device "uses" current is a somewhat loose treatment of our language, for it implies a weakening or loss of current strength. The word "use" simply means that the current performs some work. It may light a bulb; it may turn the shaft of a motor; it may be put to work for heating or cooking—these uses do not diminish the amount of current in any way.

The flow of current from a source requires two paths: one going, the other returning. That path can consist of a pair of wires, a wire and a length of ordinary metal, such as a pipe, or a single length of wire and a return path consisting of the ground. Both paths should have the same amount of resistance—something that is not true when one of the paths is a copper wire and the other path is through the ground.

The total resistance of the path consists of the resistance of the wire conductor, the resistance of the load, and the resistance of the ground if ground is being used as one of the conductors.

SOLDERLESS CONNECTORS

Wires can be joined by soldering them or, more conveniently and quickly, through the use of solderless connectors (also known as wire caps) or by split-bolt connectors.

Split-Bolt Connectors

These are commonly used for joining wires that are intended to carry heavy currents and are too thick to splice. There are various types of these connectors: some use screws for tightening the wires and holding them in place (Figure 2-1), while others

Figure 2-1   Screw-type connectors.

use a snap-in metal for locking the wires into position (Figure 2-2). The screw type is better, provided the screw does not work its way loose, something it may do if it vibrates or if a lockwasher isn't used with the screw.

Figure 2-2   Snap-in connector.

Still another member of this family is the spring-lock split-bolt, as shown in Figure 2-3. The bolt uses spring action to lock the wires in place. The snap design permits thick wires to be joined rapidly. Unfortunately, only a relatively small part of the wires' surface areas make contact, permitting a power loss.

Figure 2-3   Spring-lock connector.

Wire Nuts

Wire nuts are available in various sizes. The wire nut shown at the left in Figure 2-4 contains a threaded insert. To use this nut, join the exposed ends of the wires, twist them together, and then screw the wire nut in over the wires. The wires to be connected are made of copper that is soft enough to let the wire nut cut small threads into it, forming a screw and nut combination. Twist the plastic housing of the wire nut firmly over the joined wires until the plastic reaches beyond the insulation of the wires.

The wire nut shown at the right in the illustration has a removable insert. This insert has a setscrew for clamping and holding the two wires in position. After the wires are joined and the screw is tightened, insert the plastic shell onto the insert to cover the joint. The wire nut at the left in the drawing is the one more commonly

Figure 2-4 Types of wire nuts. At left, plastic shell is slipped over twisted-wire pair. This is the more commonly used type and is available in different sizes. Wire nut at right has removable shell to permit wires to be screwed together. Shell is then put back over joined wires.

used. There are, however, several precautions. The wire used is generally stranded. To form the equivalent of a single conductor, twirl the wires so they become similar to a single wire. This can be done by using your fingers, but a better technique is to twist the wires with the help of a pair of gas pliers. A good technique is to mount the insulated portion of the wires in a vise. Twist the twirled ends of the wires around each other, starting the job manually but finishing it with the gas pliers.

A common mistake is to take one wire and wrap it around the other. Instead, put the two exposed wires side by side and apply the twisting action to both wires simultaneously. The wire nut must cover the exposed wires completely. One advantage of using wire nuts is that the joint can be easily disconnected should it become necessary to do so. In some instances the wire nut is wrapped with plastic electricians' tape. This supplies additional insulation and also minimizes the possibility of the nut's coming off the connection. It does make it more difficult to remove the nut if you should ever wish to do so.

How to Splice Wire

It isn't always possible to use wire nuts. The alternative is to splice the wires and then to solder them. It may also be necessary to cover them with electricians' tape. Splicing the wires joins them physically, supplying mechanical strength. Soldering them supplies good electrical continuity. Both steps may be necessary. The joined wires are then covered with electricians' tape for insulation. There are various types of wire splices.

WIRE TAPS AND WIRE SPLICES

The terms *wire tap* and *wire splice* are often used interchangeably although they are different. Joining the exposed ends of a pair of wires is a *splice,* sometimes done as a means of extending the length of a wire. If the insulation is removed from a small portion of wire anywhere along its length and another wire is joined to it at right angles, the result is a tap. The original wire is not lengthened, but it does supply a pair of wires in two directions. Figure 2-5 shows the formation of a tap. Taps and splices are part of a family known as joints.

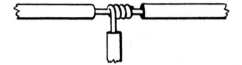

Figure 2-5 Tap joint.

The Knotted Tap

To make this tap, strip about 2 inches of insulation from the main wire to which the tap wire is to be connected. Remove about 3 inches of insulation from an end of the tap wire. Bend the tap wire over the main wire and then under it, forming a knot. Continue wrapping the tap wire around the main wire, making the turns as tight as possible. Use a pair of gas pliers, rather than fingers, to get a secure wrap. Put a small length of electricians' tape around the wrapped end of the tap to hold it in place, pending soldering the joint. The advantage of the knotted tap joint (Figure 2-6) is that it makes the connection more mechanically secure. It is still necessary to solder to get a good electrical connection.

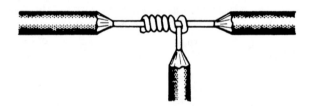

Figure 2-6 Knotted tap joint. The tap supplies mechanical strength.

The Basic Splice

Strip insulation from the ends of the wires to be joined. If the wire is stranded, use a pair of gas pliers and twirl the individual wires until they form the equivalent of a single conductor. The wires should be bare copper. If covered with enamel or any other type of insulating material, this must be removed before making the connection. The wires must be shining clean, otherwise it will be impossible to solder them.

Twist one wire around the other, as shown at the left in the illustration, Figure 2-7. Continue wrapping until the connection appears as indicated at the right. It is helpful to use a vise. Upon completion of the wrap, solder it and then tape it. Bring the joined wires into and through an electric box.

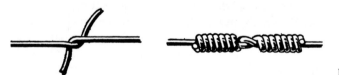

Figure 2-7 The basic splice.

The Staggered Splice

The usual method of splicing a wire to extend its length is to strip the wire ends and work from there. This does have several disadvantages. The first is that the joint can be bulky and the second is the always present possibility that the connections may short to each other. If the joint is for external wiring, such as for a lamp, it will appear unattractive. If it is to be "in wall," there may be some difficulty in getting it into a metal box.

A way of avoiding these potential problems is to use a staggered splice, as indicated in Figure 2-8. The advantages are that the two splices are offset, making the joint less bulky, and there is also the reduced possibility of a short. After making the joint, use a single wrap of tape from immediately preceding the start of the joint to its end. Staggered joints do require more time, care and patience.

Figure 2-8 Staggered splice.

Western Union Splice

While it is desirable to solder splices, this may not always be possible. Under these conditions the joint must be both mechanically and electrically secure. The Western Union splice does both.

Remove the insulation from the ends of the wires to be joined. Cross the exposed wires as shown in the drawing at the lower left in Figure 2-9. Select one of the wires and wrap it around the other. Use gas pliers to make sure this initial wrap is as tight and secure as possible. The result will be as shown in the illustration at the right. Now repeat the wire-wrapping process with the second wire, with the result as shown in the upper drawing. Both wire wraps must be tight. Check by

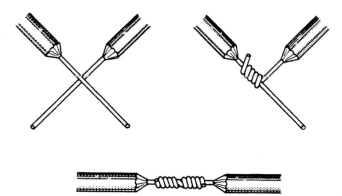

Figure 2-9 Western Union splice.

pulling on the wires or trying to rotate one while holding the other. There should be no slack or looseness when tugging on the wires in different directions. Tape the joint when it is completed.

The Rattail Joint

Also known as a pigtail joint (Figure 2-10), the rattail joint consists of a pair of stripped wire ends as shown in the illustration at the left. Twist the wires tightly around each other as shown on the right. Use gas pliers to make the joint. It will be helpful if the wire is held in a vise. After making the joint, pull on the insulated sections to see if any wire motion is possible in any direction. This is another type of joint that can be done without soldering.

Bend the joint back on either one of the conductors and then cover it with tape. Prior to making the joint, of course, the wires should be clean. If not, they can be scraped with a penknife or rubbed with a fine grade of sandpaper.

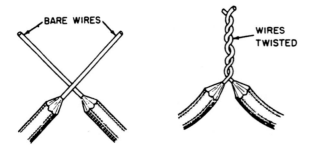

Figure 2-10 Rattail joint.

Fixture Joint

The purpose of a fixture joint is to connect the AC supply wires, a branch that ultimately returns to the fuse or circuit breaker box, to the wires supplied as part of a light fixture. The light fixture must be supported independently, often by a type of link chain or by some kind of metal housing. It is not the function of the wires to support the fixture. Quite commonly the AC supply wires will have a smaller gauge number; that is, the wires will be thicker than the wires accompanying the fixture.

There are two ways of connecting the fixtures wires to the line power wires. One method is to use wire nuts, the other is to use a fixture joint, as indicated in Figure 2-11.

In some instances the fixture wires will come with stripped ends, but if not, it will be necessary to remove the insulation. A first step is to twist the wires around each other as in Figure 2-11(a). The wire at the left is the fixture wire; that at the

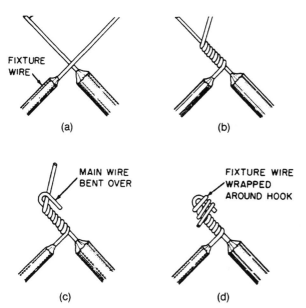

Figure 2-11   Fixture joint.

right is the branch wire or AC source wire or main wire. Do not twist the two wires around each other. Instead, wrap the fixture wire around the main wire (b). Take the main wire and bend it over the fixture wire, forming a hook, as in drawing (c). Finally, wrap the fixture wire around the hook, as in drawing (d).

There are now two options. One is to solder the joint; the other is to cover it with a wire nut. Whichever method is used, finish the job by wrapping the joint with tape. Make the wrap a tight one. Because it can stretch, vinyl tape can make a very tight covering.

Joining Three or More Wires

To join three or more wires, strip the ends of the wires to an equal length of about 1 inch. Put the three ends adjacent to each other and clamp them in a vise. Twist the lengths of wire in the same direction so they form a rope-like structure. Cover the joint with a wire nut. Always make sure that the hood of the wire nut covers the joint completely. Alternatively, solder the joint. In each case finish by using vinyl tape.

Stranded Wire Tap

Figure 2-12 shows another kind of tap, but it requires the use of stranded wire by both conductors. As a first step select the area of the wire along which the tap is to be made. Strip the insulation from the conductor as well as from one end of the

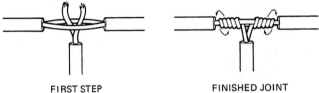

FIRST STEP   FINISHED JOINT   **Figure 2-12**   Stranded-wire tap.

wire to be joined. Separate the strands as in the conductor at the left so that it forms two groups. Each group should have approximately the same number of strands. Put the stripped end of the conductor to be joined through the space provided by the separation. After this is done, separate these strands into two groups, making certain again that there are about the same number of wire strands in each.

Twirl the exposed ends, and wrap each around the wire being tapped. Wrap one set of twirled strands clockwise, the other set counterclockwise. You will find it helpful to use pliers to make the connection as tight as possible. The finished joint can be soldered and wrapped with tape.

Alternate Method of Splicing Stranded Wire

The three steps in Figure 2-13 show an alternate method of splicing stranded wire. Loosen the strands and fan them outward. Then push the two ends together so that the strands of each will mesh, as in (a). Bend the strands of each back so that they interlock as in (b). Now twist the wires around each other, forming a somewhat solid link between the wires (c).

Precautions in Making Joints

While a joint can be made finger tight, it is always better to obtain additional force by using pliers for wire wrapping.

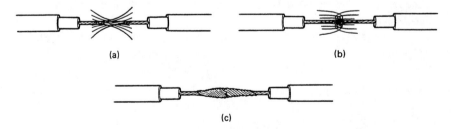

(a)   (b)

(c)

Figure 2-13   Making a stranded wire splice. (a) Fan the strands and bring them together; (b) pull the strands back, making sure they intertwine as much as possible; (c) twist the strands around each other as tightly as possible.

Try to avoid putting one joint directly above the other. The two joints will be bulky and difficult to fit into a channel, if one is to be used.

To ensure a good electrical connection, it is advisable to solder joints. Do not depend on soldering for mechanical strength.

Always tape joints to minimize the possibility of a short circuit. Use electricians' tape and make sure the wrap is tight.

When using a wire nut on a joint, make sure the apron, the open end of the nut, covers the joined wires completely. As added protection, wrap the open end with tape.

Make sure wires to be joined are clean. If necessary, use a fine grade of sandpaper, a file with fine teeth, or a knife.

When joining stranded wire, always twirl it to form the equivalent of a single, solid conductor.

If possible, use a wire stripper to remove insulation. If using a knife or diagonal cutters, cut into the insulation in a circular manner. Avoid accidental cutting or nicking any of the strands.

SCREW CONNECTIONS

Two or more wires can be combined to form a joint, but while this is a common method, there are other techniques that can be used for making a connection. Quite often, as in the case of power receptacles (outlets), a connection is made by means of a screw. The same receptacle may supply an option: screw connections or push-in terminals that grip stripped, solid wire ends automatically.

A screw connection or push-in terminals are easier to handle than making a joint, for no soldering or taping is required. A wire or wires may need to have their ends stripped, but no other preparatory work is required. However, the fact that the joining method is so simple does not mean that a correct technique can be ignored.

WIRING

Unless a home has been recently constructed, it is probable that its wiring is inadequate. It isn't easy to add more wiring to an existing installation, since much of it is hidden behind walls.

There is a mistaken conception that an unlimited source of electrical power is available through baseboard receptacles. There is no question that a local utility can meet the power needs of a home. The problem is whether existing wiring can accommodate increased electrical demands. It is possible to overload wiring just as it is possible to overwork or overload an appliance. One indication of inadequate wiring and an insufficient number of receptacles is an increased use of taps, plug-in devices for supplying more outlets.

CONDUCTORS

A conductor is any substance—gas, liquid or solid—that permits the relatively easy passage of an electric current. For home wiring the preferred conductor is copper. For a while aluminum was tried as a copper substitute, but copper wiring is preferable, since volume for volume it has better conductivity than aluminum. Copper is easily soldered. Aluminum can be soldered, but it is much more difficult to do so. In some areas the use of aluminum as an electrical conductor is forbidden.

Relative Conductivity

Silver is an excellent conductor but is expensive compared to copper. If silver is used as a reference and is assigned a conductivity of 100 percent, then by comparison copper is 98 percent, gold is 78 percent, aluminum 61 percent, platinum 17 percent, tin 9 percent, nickel 7 percent, and mercury only 1 percent.

Wire Insulation

Wire can be categorized in many ways: by whether it is a solid conductor or stranded, by its gauge number, or by the kind of insulation, if any, that is used.

The simplest kind of wire is single conductor covered with some sort of insulation, such as plastic (Figure 2-14a), cotton, silk, enamel or synthetic rubber. It can also be paper or varnished cambric. The insulation may be single layer, double or more. It may, for example, consist of an inner layer of plastic, surrounded by an outer layer of glass braid (b). The insulation, or the number of insulating layers, or both, are sometimes used as an identifying characteristic. Thus, SCC wire is single-cotton covered. DCC is double-cotton covered. The purpose of the insulation is to prevent the accidental transfer of an electric current from one wire to another, from

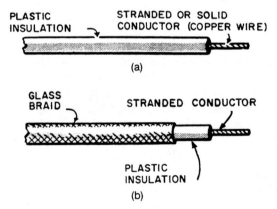

PLASTIC   STRANDED OR SOLID
INSULATION   CONDUCTOR (COPPER WIRE)

(a)

GLASS
BRAID   STRANDED CONDUCTOR

PLASTIC
INSULATION

(b)

Figure 2-14 Single and double insulated wire.

the wire to some adjacent metal such as an electric box or a wall plate, or to the housing of an electrical device.

The conductor can either be solid or stranded copper wire (Figure 2-15). If stranded, the total number of strands is equivalent in conductivity to a particular size of solid conductor.

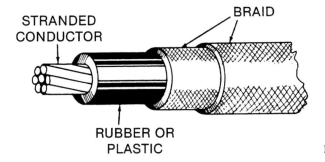

Figure 2-15   Stranded wire.

Wire Size

The size of a wire is not its length but is determined by its cross-sectional area. This is the area that is seen when looking at a wire end-on (Figure 2-16). The area of a wire is measured in a unit called a circular mil, abbreviated as CM.

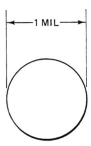

Figure 2-16   The diameter of a wire is measured in mils. The cross-sectional area is the diameter multiplied by itself.

A *circular mil* (CM) is the area of a wire whose diameter is equal to 1 mil or $\frac{1}{1000}$ inch or 0.001 inch. Wire size is specified by a gauge number usually extending from No. 0000 wire to No. 40. The thickest wire is 0000 wire and the thinnest, No. 40 (Figure 2-17).

Table 2-1, known as the American Wire Gauge (AWG), supplies data on standard annealed solid copper wire.

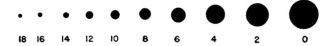

Figure 2-17   Cross-sectional areas of some solid copper wires from gauge 18 to gauge 0.

TABLE 2-1 AMERICAN WIRE GAUGE

Gauge Number	Diameter (mils)	Cross section		Ohms per 1,000 ft		Ohms per mile	Pounds per 1,000 ft
		Circular mils	Square inches	25° C (= 77° F)	65° C (= 149° F)	25° C (= 77° F)	
0000	460.0	212,000.0	0.166	0.0500	0.0577	0.264	641.0
000	410.0	168,000.0	.132	.0630	.0727	.333	508.0
00	365.0	133,000.0	.105	.0795	.0917	.420	403.0
0	325.0	106,000.0	.0829	.100	.116	.528	319.0
1	289.0	83,700.0	.0657	.126	.146	.665	253.0
2	258.0	66,400.0	.0521	.159	.184	.839	201.0
3	299.0	52,600.0	.0413	.201	.232	1.061	159.0
4	204.0	41,700.0	.0328	.253	.292	1.335	126.0
5	182.0	33,100.0	.0260	.319	.369	1.685	100.0
6	162.0	26,300.0	.0206	.403	.465	2.13	79.5
7	144.0	20,800.0	.0164	.508	.586	2.68	63.0
8	128.0	16,500.0	.0130	.641	.739	3.38	50.0
9	114.0	13,100.0	.0103	.808	.932	4.27	39.6
10	102.0	10,400.0	.00815	1.02	1.18	5.38	31.4
11	91.0	8,230.0	.00647	1.28	1.48	6.75	24.9
12	81.0	6,530.0	.00513	1.62	1.87	8.55	19.8
13	72.0	5,180.0	.00407	2.04	2.36	10.77	15.7
14	64.0	4,110.0	.00323	2.58	2.97	13.62	12.4
15	57.0	3,260.0	.00256	3.25	3.75	17.16	9.86
16	51.0	2,580.0	.00203	4.09	4.73	21.6	7.82
17	45.0	2,050.0	.00161	5.16	5.96	27.2	6.20
18	40.0	1,620.0	.00128	6.51	7.51	34.4	4.92
19	36.0	1,290.0	.00101	8.21	9.48	43.3	3.90
20	32.0	1,020.0	.000802	10.4	11.9	54.9	3.09
21	28.5	810.0	.000636	13.1	15.1	69.1	2.45
22	25.3	642.0	.000505	16.5	19.0	87.1	1.94
23	22.6	509.0	.000400	20.8	24.0	109.8	1.54
24	20.1	404.0	.000317	26.2	30.2	138.3	1.22
25	17.9	320.0	.000252	33.0	38.1	174.1	0.970
26	15.9	254.0	.000200	41.6	48.0	220.0	0.769
27	14.2	202.0	.000158	52.5	60.6	277.0	0.610
28	12.6	160.0	.000126	66.2	76.4	350.0	0.484
29	11.3	127.0	.0000995	83.4	96.3	440.0	0.384
30	10.0	101.0	.0000789	105.0	121.0	554.0	0.304
31	8.9	79.7	.0000626	133.0	153.0	702.0	0.241
32	8.0	63.2	.0000496	167.0	193.0	882.0	0.191
33	7.1	50.1	.0000394	211.0	243.0	1,114.0	0.152
34	6.3	39.8	.0000312	266.0	307.0	1,404.0	0.120
35	5.6	31.5	.0000248	335.0	387.0	1,769.0	0.0954
36	5.0	25.0	.0000196	423.0	488.0	2,230.0	0.0757
37	4.5	19.8	.0000156	533.0	616.0	2,810.0	0.0600
38	4.0	15.7	.0000123	673.0	776.0	3,550.0	0.0476
39	3.5	12.5	.0000098	848.0	979.0	4,480.0	0.0377
40	3.1	9.9	.0000078	1,070.0	1,230.0	5,650.0	0.0299

Standard annealed solid copper-wire table, using American Wire Gauge (AWG). Another guage, Browne & Sharp (B & S), is identical to AWG.

42

The first column in the table is the gauge number. The smaller the gauge number, the thicker the wire. The thickest wire is No. 0000 and has a diameter of 460.0 mils or almost a half inch.

The thinnest wire is No. 40, and its diameter is 3.1 mils or a little more than three one-thousandths of an inch. The third column is the cross-sectional area in CM. Note how rapidly the area increases as the gauge number becomes smaller. The area in CM of No. 40 wire is only 9.9 while that of No. 0000 wire is 212,000.

By examining the last column, it is possible to get some idea of the weight of the wire in its uninsulated form. A thousand feet of No. 40 wire weighs 0.0299 pound and a thousand feet of No. 0000 wire weighs 641 pounds. However, neither No. 40 wire nor No. 0000 is used in the home. For home wiring No. 14 gauge is quite common, while you will find Nos. 16 and 18 used in lines connected to various appliances. Service wires coming into the home are about No. 6 or No. 8.

While the table shows wires only up to No. 40, wire sizes all the way up to No. 60 are available. Such very fine wires, a few thinner than a human hair, are used in the wiring of some electronic instruments.

Solid wire is used for branch power lines, the lines that connect the fuse or circuit-breaker box with receptacles throughout a home. Solid wires are easier to connect to receptacles and to feed through conduits intended to hold the wire. Stranded wires are more flexible and are suitable for uses such as lamp cords and extension cables.

If a single number is specified for a wire, then it is a solid type. Nos. 12, 16 or 18 are examples of solid wires. Stranded wire is always referred to by a pair of numbers such as 18/30. This means that this wire consists of 18 strands of No. 30 wire.

Like solid wires, stranded types have current limitations. 65 strands of No. 34 wire will carry the same amount of current as a solid No. 16 wire, so for wiring purposes use either 65 strands of No. 34 or No. 16 solid. Some manufacturers will identify the wire as No. 16 stranded instead of referring to it as 65/34. A designation such as No. 16 stranded does not supply information about the number of strands nor the gauge of the individual wires.

Calculating the Solid-Conductor Equivalent of Stranded Wire

Assume a stranded wire consisting of 37 strands, each of which is 0.002″ (or 2 mils) thick (Figure 2-18). $2 \times 2 = 4$ so each strand has an area of 4 CM. Since there are 37 strands, the total cross-sectional area is $37 \times 4 = 148$ circular mils.

Look in the wire table, and under the heading of "Cross Section," in the third column, locate the areas in circular mils. Move down this column, and you will see there is no single, solid conductor that has an area of 148 CM. But No. 29 wire is 127 CM and No. 28 wire is 160 CM, so the stranded wire would be considered somewhere between No. 28 and No. 29 gauge.

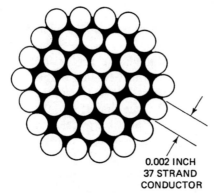

0.002 INCH
37 STRAND
CONDUCTOR

Figure 2-18 Cross-sectional view of wire made of 37 strands of 2-mil wire.

Wire Gauge

Without considerable experience it is difficult to estimate the gauge of a wire simply by inspection. If the wire is on a spool, the gauge will be indicated somewhere on one of the side supports. The wire gauge may also be printed on the wire insulation.

The wire gauge in Figure 2-19 is a useful tool for measuring wire sizes. To use the gauge, remove insulation from one end of the wire. Select a hole in the wire gauge that looks as though it will allow the wire to pass through it. If it does so very easily, try the next smaller hole. Continue until you find the one hole that will permit the wire to pass through snugly.

The wire gauge has each hole marked with wire size data. On one side the diameter of the wire is supplied in decimal fractions. Using the same hole the AWG gauge number is on the opposite side.

The greater the cross-sectional area of a wire, the more current it can carry. At the same time it is more expensive than thinner wire and is more difficult to

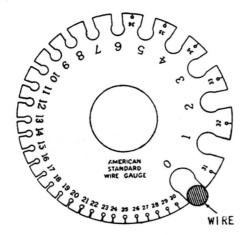

WIRE   **Figure 2-19** Wire gauge.

handle in making connections. Cross-sectional area isn't the only factor in determining current-carrying capacity. It is also affected by the length of the wire, temperature, and the insulation surrounding the wire.

The diameter in mils shown in column two in the American Wire Gauge table has an accuracy to one decimal place. These figures are reasonable working numbers but aren't absolutely precise. For this reason, squaring them, multiplying each diameter by itself to obtain the cross-sectional area in circular mils, will not supply results which correspond to the data in column 3.

As an example consider the listing for gauge No. 40 wire at the bottom of the table. The diameter is supplied as 3.1 mils. Multiplying 3.1 by 3.1 is 9.61 CM and not 9.9 CM as shown. The reason for this is that the diameter of gauge No. 40 wire is not simply 3.1 mils but 3.1464266 mils. For practical electrical wiring, 3.1 mils is sufficiently accurate and so is the area in CM shown as 9.9.

Resistance of Wire

The smaller the cross-section of a wire, the greater is its resistance per unit length. Resistance is the opposition to the flow of current and is measured in ohms. A foot of No. 12 wire, for example, has more resistance than a foot of No. 8 wire. Whether resistance is important in home-wiring calculations depends on wire length and on how much current is to flow through the wire. The amount of current, in amperes, depends on the load. The greater the load, the heavier the demand for current.

There are two possibilities when using the wrong gauge wire for current-hungry appliances. Because of the resistance there will be a voltage loss, a voltage drop, along the wire. Further, the wire may become hot. Assume that 10 amperes is to flow through a wire and that the length of the wire between the source and the appliance has a total resistance of 1 ohm. To find the voltage loss in the wire, multiply the current flow by the resistance of the wire. The current is 10 amperes, the resistance is 1 ohm. $10 \times 1 = 10$ volts. This means the appliance will receive 10 volts less than it should. If the source voltage at the fuse or circuit breaker box is 120 volts, then it will be 10 volts less at the input to the appliance, or 110 volts. Whether or not the appliance will work well with just 110 volts is a guess. It may or may not. Or it may not work at all.

Selecting a Wire

When choosing a wire size for installation, take two factors into consideration: how much current the wire will be required to handle and how many additional appliances will be used to load the line, either presently or in the future. The safest procedure is always to assume a worst-case situation and to use a wire having a smaller gauge. However, the smaller the gauge, the more costly the wire and the more difficult it will be to handle—to make connections.

Determination of the Total Voltage Drop

It isn't enough to select a wire based on current capacity alone. The distance from the main fuse or circuit-breaker box to the appliance is also important, particularly if the distance is long and the current demands are heavy. Generally, the allowable voltage drop in a wire is 3 percent. The actual voltage at an appliance is the voltage measured across its input terminals with the appliance operating.

To calculate the total voltage drop along a branch power line, locate the electrical appliance at the end of the line. Turn on all of the power-operated devices connected to the branch, then measure the voltage at the input to the appliance at the end of the run. The junction box (also known as an electric box) at the end of the run will not contain wires that continue on to other receptacles.

In a properly operating branch power line, there will always be voltage present at each of the receptacles. If all the appliances connected to the receptacles are turned off, the voltage that is present will be the same as that at the fuse or circuit-breaker box. It is only when appliances are working that there will be a voltage drop in the branch power line.

WIRE TYPES

There are numerous types of wire and several different kinds may be used in the home. Wire is identified by whether it is solid or stranded, its gauge, and by the kind of insulation used, if any.

The letter T is used to designate wiring for in-home use and is an abbreviation for thermoplastic insulation. Do not confuse it with TW, intended for outdoor use with possible damp or wet locations. Still another type is NW, also for indoor use and for dry conditions only.

SPT is the designation for lamp cord, HPN or HPD for heater cord.

Metallic Armor

The metallic shield covering some wires can be wire braid in which individual wires used for making the braid are woven. The braid can be made of steel, copper, bronze, or aluminum (Figure 2-20). Another type is made of steel tape which is then wrapped around the cable and covered with a layer of jute. (Figure 2-21)

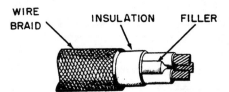

Figure 2-20 Wire-braid-type armored cable.

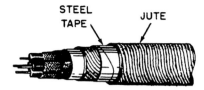

Figure 2-21 Cable enclosed by steel tape.

Wire Identification

Wires are often designated by letters rather than the more elaborate process of describing them in detail. Rubber-covered wire is called type R, and a better grade is type RH because of its heat resistance. Thermoplastic materials, with the letter T as the designation, are now used extensively. The letters NM and NMC are used for nonmetallic sheathed cable.

Lead Sheathed Cable

Lines used for bringing power into the home may be lead sheathed. These lines are owned, installed, and operated by local power companies and are not subject to servicing by the home owner.

Home Wiring

There are several different types of wire used in the home. One of these is two-wire lamp cord (Figure 2-22), which commonly accompanies appliances such as lamps, electric shavers, fans, and so on. This is plastic- or rubber- or synthetic-rubber-covered stranded wire, selected for flexibility. The accompanying plugs are molded and can be regarded as integral with the wires. A heavier gauge, but along the same lines, is used for broilers, television sets and air conditioners. Older appliances used two-prong plugs; more modern versions have a three-prong unit, with one of the prongs made specifically for grounding.

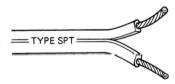

Figure 2-22 Lamp cord is designated as type SPT. The wires are stranded.

BX

For the home BX (Figure 2-23) is a type of armor once used more commonly. Where there are conditions of moisture, there is a type of BX in which the armor is put over a lead sheath. BX is found in older homes, for it has been largely replaced by

Figure 2-23 BX cable. Note absence of ground wire, subsequently included.

Romex, unshielded wire having a tough plastic-type outer covering. For home use Romex consists of two or three active wires color coded black and white, or black, white and red. Each of these has another, individual, wire for making ground connections. The black and red leads are the hot leads, the white is neutral. The ground wire is either bare or is insulated with a green color-coded covering.

BX, a flexible, metallic-covered cable, can be regarded as intermediate between Romex and conduit. Given a choice, Romex is much easier to handle. BX is very difficult to install once walls are in position, but it can easily be used in basements or in attics having exposed beams.

The advantage of BX is its metallic shield, at one time used in place of a ground wire. Unlike conduit, it isn't necessary to fish wires through it, and BX does not require a special bending tool. When buying BX, make sure it has a ground wire running through it. Just because BX can be bent does not mean there are no limits. Don't try to square off this cable; that is, do not try to make sharp right-angle turns. Make all turns as gradual as possible to avoid damaging the armor (Figure 2-24). Once adjacent turns of the armor separate, there is no way to get them back to their original position. The illustration also shows a through hole drilled in one of the studs for supporting the cable. The hole should have a diameter at least ⅛" larger than the outside diameter of the BX to permit pulling the BX through.

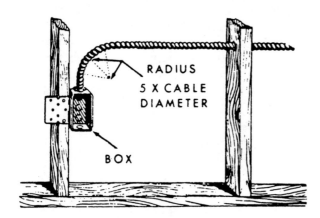

RADIUS
5 X CABLE
DIAMETER

BOX

Figure 2-24 Avoid sharp bends when installing BX.

When running BX along a beam, use staples of the kind shown in Figure 2-25. The installation will be easier if the box connections are made first but support the cable with one or more staples along the length of its run. Use staples every 4½ inches, but if the BX shows a tendency to sag, change the support separation to 2 inches. As a final step the cable should have a support staple not more than one to

Figure 2-25 Offset staple used to anchor BX.

two inches away from the box. The reason is to avoid any mechanical strain on splices or connections inside the box. One of the advantages of BX is it is very unlikely (unlike Romex) that hammering staples can damage the cable.

How to Prepare and Attach BX

One of the great disadvantages of BX compared to Romex is the difficulty of cutting BX. Romex can be cut with a pair of diagonal cutters or a hacksaw used at right angles to the cable. The easiest and fastest way of cutting BX is a special tool designed for that purpose. The alternative is to work with a hacksaw.

When planning to cut with a hacksaw, take two preliminary steps. Use a new blade or one that is reasonably new. Some blades are better and more durable than others, have sharper teeth, and are designed to cut through metal.

When inserting the blade into the hacksaw, make sure the teeth point away from the handle. Also make sure the blade is tight and does not bend or curve along its length. The hacksaw is equipped with a thumb screw to permit making this adjustment. Hacksaw blades can and do snap if not inserted or used properly.

As a first step, measure the length of BX that will be needed. Include all bends and make some allowance for the fact that the BX may not follow an absolutely straight line when installed. It is possible to cut BX while holding it, but this is both awkward and risky. Use a vise and then cut at right angles to the cable.

Once the cable has been cut through, use the hacksaw again, but this time just to remove a section of the armor to expose the wires that are inside. Cut the armor about 8 inches from the end (Figure 2-26) so there will be ample wire to go inside an electric box. It will be necessary to cut very carefully to avoid damaging the wires inside the armor. To do this, put the cable into a vise with an 8-inch length extending

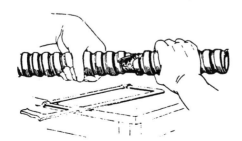

Figure 2-26 Once the BX has been cut to the right length, hacksaw a section of the metal shield about 8 inches from the end. Be careful not to cut the wires.

from its jaws. This time cut at right angles to the cable with the blade of the hacksaw right up against the jaws of the vise, using the edges of the vise as a guide. Do not cut continually at one spot but move the blade so it follows a semicircle. Make the strokes light and do not exert too much pressure. You will be able to tell when the blade begins to get through the armor since it will then tend to grip the blade, making it more difficult to move. Now loosen the vise and rotate the armor. Continue cutting again in semicircular form.

When you finish, you will have cut a circle around the armor. Gently keep deepening the cut. Finally, it will be so deep you will be able to bend the cable back and forth, breaking the armor at the cut. If some of the armor still remains unbroken, cut it away with the hacksaw.

After the armor has been cut, pull it off carefully to avoid damaging the enclosed wires. Unwrap any paper enclosing the wires and cut it away.

This completes cutting the BX at one end. Do the same job with the other end. When you have finished, some of the edges of the BX may be rough. File these with a flat file, but be careful not to damage the insulation of the wires. Insert an antishorting bushing (Figure 2-27) by sliding it over the wires, then insert the bushing so it fits snugly between the wires and the inside of the BX armor.

Figure 2-27   Insert insulating brushing between wires and armor to keep sharp edges of armor from cutting into wires.

CONDUIT

Conduit is hollow pipe generally ½ inch or ¾ inch in diameter. Its purpose is to protect the wiring it encloses, and it is superior in this respect to BX cable or Romex. Romex, however, is easier to handle and does not require special bending tools, while BX is difficult to cut, and its use may not be permitted by local electrical codes.

Conduit is available in three different forms: plastic, thin-wall metal, and rigid threaded. Thin-wall conduit costs less than rigid and instead of having threaded fittings is a force-fit type. Because of its structure it is also easier to bend.

Before replacing any wiring, check local electrical codes or consult a licensed electrician. Your fire insurance policy may be affected by the type of wire housing you select. Rigid and thin-wall conduit can be used either indoors or outside, but rigid conduit is preferable for underground work. Again, there may be local restrictions on how you may use rigid conduit underground. While both types of conduit, rigid and thin wall, may be galvanized, some soils and building materials may be too corrosive to permit their use.

After conduit is installed, the required wires are snaked or fished through.

Thin-wall conduit is the preferred type for in-wall wiring. It is sold in 10-foot lengths with the ½-inch type (outside diameter or OD) capable of containing four gauge No. 14 wires or three gauge No. 12. Thin-wall ¾-inch will accept four gauge No. 10 or five gauge No. 12 wires. The wires that are used are individual, not paired, types.

PLANNING A RUN WITH CONDUIT

Assume you want to install an electric box somewhere near the fuse or circuit-breaker box in your basement. Your first problem is to decide whether to use thin-walled or rigid conduit or Romex. With rigid conduit you will use threaded fittings, so connecting the conduit at either end will be easier. This means you will need to buy a die for cutting the conduit threads unless you can buy threaded conduit cut to the size you require. If you use thin-walled conduit, you will find it easier to bend, and you will be able to use compression-type fittings.

While most thin-walled conduit is ½ inch or ¾ inch in diameter, it is possible to obtain it with other dimensions. Table 2-2 lists the number of wires in one conduit for various wire sizes.

TABLE 2-2   NUMBER OF WIRES IN ONE CONDUIT

Avg Size Wire	Minimum Size Conduit Permitted							
	1	2	3	4	5	6	7	8
14	½″	½″	½″	½″	¾″	¾″	1″	1″
12	½″	½″	½″	¾″	¾″	1″	1″	1″
10	½″	¾″	¾″	¾″	1″	1″	1″	1¼″
8	½″	¾″	¾″	1″	1¼″	1¼″	1¼″	1½″
6	½″	1″	1″	1¼″	1½″	1½″	2″	2″
4	½″	1¼″	1¼″	1½″	1½″	2″	2″	2″
2	¾″	1¼″	1¼″	2″	2″	2″	2½″	2½″
1	¾″	1½″	1½″	2″	2½″	2½″	2½″	3″
0 . . . . . .1″		1½″	2″	2″	2½″	2½″	3″	3″
00 . . . . . .1″		2″	2″	2½″	2½″	3″	3″	3″

The next step is to determine the location of the electric box and its distance to the fuse or breaker box. Take into consideration any conduit bends that may be needed. You may need to buy or rent a conduit bender. Don't try to bend conduit by hand or by putting the conduit in a vise. The bends must be gentle. Figure 2-28 shows that the conduit has two bends and that these are actually 90 degrees.

Before fastening the conduit into position, mount a locknut at each end. Insert one end into the fuse or breaker service box and the other end into the juction box.

Mount the conduit against joists or other wooden supports. Put a locknut over that section of conduit that extends into the power service box. Fasten the junction

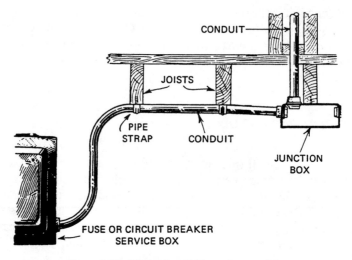

Figure 2-28   Wiring installation using conduit.

box into position by using wood screws going into a wooden support behind the box. As a final step, tighten all locknuts. There will be two at the service box (one inside, one outside). If necessary, add more pipe straps to support the conduit.

Figure 2-29 shows the details of a thin-wall conduit installation. Thin wall connects only to a steel electric box, never to plastic types. The conduit is available in a number of lengths but is easy to cut to an exact size with a hacksaw. After cutting, ream the ends on the inside and then taper them with a file. Use a conduit bender to make bends. Force a connector over the end of the conduit, insert into an electric box and then tighten with a locknut. The coupling shown in the illustration is used to connect one section of thin-wall conduit to another.

With the conduit now fixed into position, fish the wires through it. You know the length of the conduit, so cut the wires accordingly. It's a good idea to allow about 6 inches of extra wire length at each end. If the conduit is relatively short and if the bends are gradual, you may be able to push the wires through. Before trying it, make sure that the wires are straight with no kinks and that the ends of the wires being pushed into the conduit aren't knotted or curved.

Should you fish the wires through before or after putting the conduit into position? That is up to you. With the conduit in place you will have eliminated one item that needs to be held. On the other hand, the conduit may be in a location that makes inserting the wires rather difficult. Do what is easier for you.

Making the Connections

The conduit is now installed, and the wires have been fished through it. If you plan to use an electric box as an active power receptacle, all you will need to do will be to connect the wires to the receptacle. Alternatively, you may decide to use the box

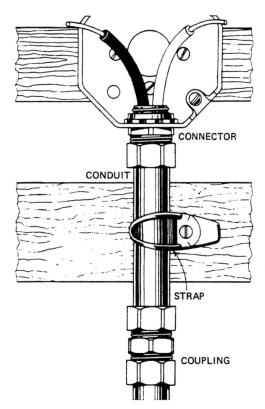

Figure 2-29 Installation of thin-wall conduit. (Courtesy Sears Roebuck & Co.).

as a junction type for future use. In that case cover each wire with a solderless connector, then put a plate over the box and screw it into position. Unlike wall plates used for receptacles or switches, this plate will have no holes or cutouts in it, other than those used for mounting screws.

Knockouts

It is easy, particularly for those with limited experience, to drive out more knockouts than they really need. The use of excessive knockouts is an indication of lack of wiring planning.

A knockout consists of a circular bit of metal that can easily be pushed out of an electric box. If you do have one or more knockout holes and you do not use them for entering or exiting cable, then you must cover them. Do this by inserting knockout covers.

There are a number of reasons why knockout holes should be covered. In the case of a fire inside the box, it will be confined to that space. While the covered knockout hole doesn't make the electric box airtight, it does limit air access, so any interior box fire will be limited to that extent. Knockout covers also protect against water leakage, dust, and fumes.

The drawing in Figure 2-30 shows how conduit is continued from a basement to the first floor. The conduit is concealed behind walls. It is possible to add more boxes for switches and receptacles along the length of the conduit by adding the conduit in sections.

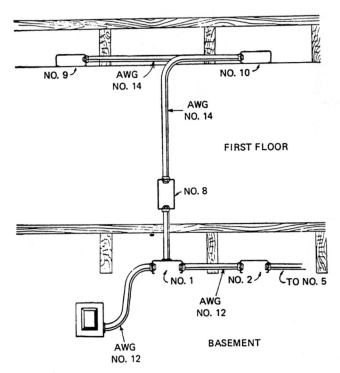

Figure 2-30 Make a written plan, as shown here, and identify each electric box by number. On the plan indicate the size of the wire used inside the conduit. Adequate wiring and planning are helpful when selling a house.

Wire Pulling

Wiring should be pulled into the conduit after electric boxes have been installed. If the conduit is short enough, the wires can be paired and pushed through the conduit from box to box. If the conduit run is longer and has several bends and more than two conductors, a fish wire (Figure 2-31) can be used in pulling the wires through. The fish wire has a hook shape at one end. Make a corresponding hook out of the wire ends and join the two. Another method is twisting the leads of the wires around the fish, and sometimes using electrical tape will be helpful.

Pulling wire through conduit becomes easier if it is made into a two-person job, with one pulling on the fish wire and the other feeding the wires into the conduit. For a long run it may be helpful to cut the wires after they have been passed

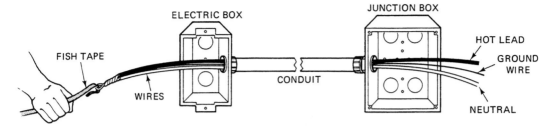

Figure 2-31 Method of fishing wires through conduit.

into an electric box. Wire splices can be made in electric boxes, but avoid making any splices inside the conduit.

Supporting Conduit

Conduit is supported about every four to six feet by a pipe strap. The straps are screwed into studs. If a length of conduit ends somewhere between electric boxes, extend it by using a coupling. Special cutters are made for thin-wall conduit, so it is quite easy to cut the conduit to a desired length.

While thin-wall conduit can be made to have a right-angle bend, this may not always be required and may be controlled by the positioning of an electric box, as indicated in Figure 2-32. Here the conduit is brought into the left side of the box. The wires are brought through the conduit and are allowed to extend to a length of about six inches and are then cut. The next length of conduit is connected through a knockout at the bottom center of the box. The conduit is fastened to the box and to the following box as well. The wires for this conduit are pulled through it and then are also cut to a length of about six inches. The wires are then stripped and joined, black to black and white to white. The ground wires, not shown in the illustration, are stripped and fastened with screws to the box.

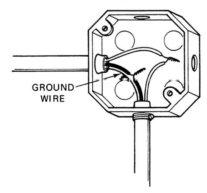

Figure 2-32 Connection of conduit to electric (junction) box.

Additional Wiring for Conduit

If additional wiring is to be installed, determine if the existing conduit is capable of carrying it. It may be possible to fish through the conduit connected to an existing box to bring the new wires almost all the way to the location of a new box. Being able to do this will save work and expense.

Intermixing

The fact that an older home has in-wall wiring using conduit does not prevent the use of a plastic-sheathed cable such as Romex when wiring updating is needed.

Couplings and Box Connections

Couplings and fittings (Figure 2-33) to electric boxes can be clamp or compression types. Some fittings are similar to sleeves and can be fastened to conduit by an impinger tool. This tool pinches circular indentations in the fitting which holds it firmly against the conduit. Others have threaded bushings that, when tightened, force a tapered sleeve firmly against the tubing. (Figure 2-34).

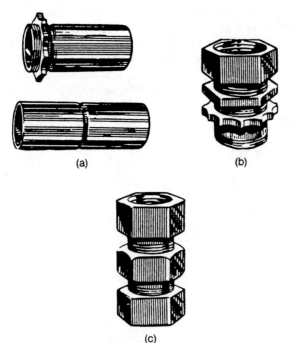

(a)

(b)

(c)

Figure 2-33 (a) Indent-type fittings for thin-walled conduit; (b) box connector for thin-walled conduit; (c) coupling for thin-walled conduit.

Figure 2-34 Couplings and connectors for flexible metal conduit. (a) Unit for connecting two sections; (b) coupling for joining rigid to flexible conduit; (c) connector for joining flexible conduit to a metal box.

(a)   (b)   (c)

Wire color coding must be followed when using conduit: black and red for hot leads, white for neutral, and green or bare for ground leads. When counting wire sets, the ground wire is not included. Thus a two-wire set would consist of one black and one white. The green or bare ground wire is not part of the count. A three-wire set includes a black, red, and white, with the green or bare wire not counted.

Conduit cannot be snaked through a wall in the same manner as plastic sheathed cable since it does not have the cable's flexibility. It can be used in areas where the house framing may be exposed, as in basements or attics, or in living areas where a wall is being replaced or has not yet been installed.

While the dimensions of conduit are the same as those used by some water pipes, one is not a substitute for the other. The inside conduit finish is smoother, so the insulation of wires will not be abraded as the wire is fished through. There are some wires whose outer insulation is especially prepared to facilitate fishing through rigid conduit. The treatment consists in giving the insulation a wax coating so that wires can be made to slide more easily against each other or against the inner walls of the conduit.

Conduit is softer than water pipe, so it is more easily bent. Some conduit is available in aluminum, copper alloy or a plastic-jacket for installation where the atmosphere is corrosive.

Conduit Benders

Thin-wall conduit can be bent with a special tool made for the purpose. The conduit can be shaped to any angle up to 90 degrees as shown in Figure 2-35. However, the greater the angle and the more bends made in the conduit, the more difficult it will be to fish wires through.

Rigid Conduit Installation

The finish on rigid conduit is either black enamel or galvanized. Galvanized conduit is generally approved for indoor or outdoor installation, while the black enamel type is restricted to in-house use.

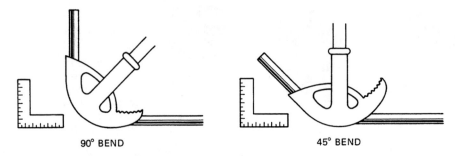

90° BEND   45° BEND

Figure 2-35 Conduit can be bent to any angle up to 90 degrees.

Rigid conduit has the same size designations as water pipe. The size of the conduit is specified by its inside diameter (ID). While ½ inch is a commonly used size, some $\frac{5}{16}$-inch conduit is permitted where extensions are to be made under plaster. ½-inch conduit has an actual inside diameter of 0.622 inch. Standard conduit sizes used in interior wiring are ½, ¾, 1-¼, 2, and 2-½ inches, but sizes up to 6 inches are available for commercial installations.

ROMEX

The problems with conduit, whether thin walled or rigid, are the difficulty in handling, the need for fishing wires through it, and bending. For indoor installation, for in-wall and out-of-wall such as exposed basement wiring, it may be permissible to use Romex (Figure 2-36). This nonmetallic sheathed cable has become so widely used that the trade name, Romex, is sometimes applied to all cables of this type.

Nonmetallic sheathed cable has the disadvantage that improper handling can damage the wire. The insulation can be damaged by hammer blows.

Romex NM-B cable and UF cable are available with or without grounding conductors. The wire comes in three sizes: 14 AWG, 12 AWG and 10 AWG. 14 AWG is the minimum size recommended for general purpose lighting circuits; 12 AWG is intended for 20-ampere circuits such as kitchen or small appliance connections. 10 AWG is the minimum size for 30-ampere circuits such as those supplying power to washers, dryers, and electric ranges.

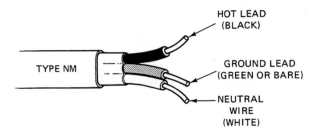

HOT LEAD
(BLACK)

TYPE NM

GROUND LEAD
(GREEN OR BARE)

NEUTRAL
WIRE
(WHITE)

Figure 2-36 Romex cable. The designated code is NM and the wire is intended only for dry, indoor installations.

For indoor applications either NM-B or UF cable can be used. Romex NM-B has a tough ivory colored flexible outer jacket placed over individually insulated conductors. Romex UF is gray colored and is moisture, sunlight and corrosion resistant for direct burial. The construction is a solid thermoplastic jacket over individually insulated copper conductors. Romex UF is available with or without a ground conductor, but when installed outdoors use only Romex UF equipped with a ground cable.

NM and NMC Cable

NM (nonmetallic) and NMC (nonmetallic cable) are two of the family of nonmetallic cables. NM is for indoor use only, while NMC is intended for either indoor or outdoor work. The sheath or protective covering of NMC is tough, much more so than NM, although it isn't possible to determine this by making a side-by-side comparison.

NMC is available in various grades, not only for outdoor work, but for underground use or for stringing through the inside openings of concrete blocks.

UF and USE Cable

Other types of non-metallic cable are UF (underground fused) and USE (underground service entrance) types.

EXTENSION CORDS

These are used for making temporary connections between a receptacle and some appliance. They extend the fixed receptacle to a point some distance away and are available in lengths from 3 to 100 feet. In its basic form the cord consists of a two- or three-prong plug at one end and a receptacle at the other. The plug is inserted into a live receptacle, while the other end of the extension cord is used to accommodate the plug of some appliance. The extension cord's receptacle can be a single type, a duplex and, in some instances, a three-plug type.

Some extension cords are equipped with a light, sometimes called a drop light. The drop light may have its own receptacle. At one time extension cords had two wires only and used two prong nonpolarized plugs. The modern cord is a three-wire type and uses polarized plugs only. The three wires consist of a hot lead, a neutral, and a ground wire.

When an extension cord is plugged into a live receptacle, it becomes part of the house wiring. The amount of current passed by the cord cannot exceed the current limitations of the wiring branch line to which it is connected.

To select the right extension cord, check the current rating of the appliance, then choose the correct size and length extension cord. The current rating of the cord is usually indicated on its packaging.

If the selected cord has a current capacity exceeding that of the appliance to be used, there is no difficulty. However, since the current taken by the appliance must flow through the cord, there will be a voltage drop, and this will be across the entire length of the cord. This reduces the amount of voltage available at the appliance. Consequently, there are two rules that are applicable to extension cords. It is desirable, but not always practical or convenient, to use a cord that is no longer than necessary. Another factor is to use a cord having a better current rating than the appliance to which it will be connected.

After using an extension cord, remove its plug from the receptacle to which it was connected and hold the prongs. If they feel hot to the touch, the current-carrying capacity of the cord is inadequate for the particular appliance being used, although it may be perfectly satisfactory for some other, less current-hungry appliance.

Extension cords are identified with the code name SJ. A typical example of an extension cord would be one made of three No. 16/30 conductors (16 conductors of No. 30 wire) and capable of carrying 7 amperes. This cable is equivalent to three No. 18 solid copper conductors. Stranded wire is used for extension cords to supply flexibility.

Choosing the Right Cord

Cords can be categorized as medium, heavy and super. A medium cord will use the equivalent of 16-gauge wire and will have a current capacity of 10 to 13 amps. Typical uses are for a popcorn maker, a coffee pot or a waffle iron. A heavy cord will have 14 gauge wire and a current capacity of 13 to 15 amps. It will work with a small refrigerator, a room air conditioner and a dehumidifier. The super uses 12-gauge wire and will pass a current of 15 to 20 amps. Typical uses are for a large air conditioner, a refrigerator or a portable heater.

Extension Cord Safety

Safety is involved in the use of extension cords just as it is with every other aspect of electricity. For your protection:

- Uncoil the extension cord before using it.
- Make sure its plug is fully inserted into the power receptacle.
- Keep children and pets away from the cord. Do not let children use the cord as a jumping rope or other plaything. Do not let children plug the cord into a receptacle.

- Keep the cord away from water. Do not drag the cord along the ground. Carry it.
- Do not attempt to splice the cord, adapt it in any way, repair or modify it. If the cord is cracked, cut, worn out or frayed, or if its plug has been damaged, dispose of it.
- Inspect the cord periodically for damage. If you do find damage, get a new cord.
- An extension cord is a temporary device. Never use it as a permanent power source. Do not use the cord overnight for operating appliances.
- Make sure no one can trip over the cord while it is being used.

LAMP AND POWER CORD

Although commonly referred to as lamp cord, this wire type is also used for fans, radios and small appliances. The wires are covered with a plastic insulating material. The conductors consist of 41 strands of No. 34 wire, the equivalent of No. 18 solid. This cord has a current rating of 7 amperes. It has a polarity ridge over one conductor, helpful in applications in which it becomes necessary to identify one of the two conductors.

The cord is rip-type construction, meaning you can separate the conductors just by holding them and pulling apart. The cord will then split down the center of the insulating material, facilitating connections with the individual wires. The identification letters for this type of lamp and power cord are SPT.

WIRE INSULATION

All house wiring, whether behind the wall or exposed or outdoors, must be insulated. Insulation may be a minimum and may consist of nothing more than a covering of rubber or a rubber-like material over copper wires, or it may include several layers of insulating materials. The greater the amount of insulation, the more rigid the wire and the more expensive it becomes. Lamp cord is extremely flexible, but carries the minimum amount of insulation.

Plastic Insulation

A large variety of different plastics are used as insulating materials for wires. Natural rubber is a vegetable product and so ultimately decays. In time rubber becomes hard and then cracks. Wire covered with aged rubber is a fire hazard. Plastic insulation has excellent flexibility, is moisture resistant, and, depending on the type of plastic can have a longer useful life than rubber.

Varnished Cambric

Whenever a current flows through a wire, heat is produced. Cambric has little insulation value when used alone. When impregnated with high-grade mineral oil, it works well as an insulating material for high-voltage cables. Paper insulation is sometimes used for underground service conductors.

Cotton and Silk

Cotton is commonly used as an insulating material, although silk was once very popular. Either of these may consist of a single or double wrap. SCC wire is single cotton covered. DCC is double cotton covered. SSC is single silk covered and DSC is double silk covered.

Enamel

This is an insulating material whose thickness depends on the number of times the wire is coated during its manufacture. One of the advantages of enamel is that it is the thinnest insulation that can be put on wires. However, the enamel has a tendency to crack when the wire is flexed. It is sometimes used as a primary insulation material, but it is never used alone. Instead, it is covered by some other insulating substance.

It isn't possible to solder directly to enamel-covered wire. The enamel must first be removed from the soldering area. Similarly, if the wire is to be connected to a screw-type terminal, it is important to remove the enamel before wrapping it around the screw. Failure to do so means either no electrical contact or an extremely poor one.

There are also various synthetic enamels that look like enamel but do not have its shortcomings. These materials have various trade names. They do not crack when flexed, are much better insulators than enamel, but are very difficult to remove prior to making connections.

Wire Protection

Despite the fact that wires are generally fixed in position, they are subject to outside effects that lower their useful life. Moisture, high temperatures, polluted air or contact with other materials all contribute some form of corrosion.

For these reasons wires are protected by some form of covering—some type of insulating material such as plastic. Over this may be an outer covering of synthetic rubber or its equivalent, which in turn may also be covered.

SOLDERING

The purpose of soldering is to join two metals electrically, usually copper wires. Soldering is not intended to make a wire connection mechanically secure (Figure 2-37).

The solder that is used is an alloy known as soft solder. In soft soldering, one of the metals—the solder—melts with the help of a soldering iron, although other heating devices can be used. The copper wires to be joined do not melt, for the heat is never high enough to reach the melting point of this metal.

There is another soldering process known as brazing or hard soldering in which silver solders or brazing alloys are used. The temperature is much higher than that in soft soldering, and the metals to be joined actually fuse. For electrical work soft, rather than hard, soldering is used.

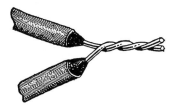

Figure 2-37 Before soldering, twist wires to form a good mechanical joint. Wires must be shiny clean over entire exposed surface area. Clean shortly before soldering as copper wire oxidizes. The oxide makes soldering difficult.

Soft Solders

Soft solders are made of tin and lead in some proportion but may also contain very small amounts of antimony, bismuth, cadmium, and zinc. The quantity could be about 0.1 percent bismuth and 0.1 percent antimony, so electrical solder is primarily a tin and lead combination.

Eutectic Point

Lead has a melting point of 620°F while that of tin is 450°F. However, when these two metals are combined, the lowest melting point obtainable is with an alloy of about 63 percent tin and 37 percent lead. This combination of metals is called eutectic, and the melting point is called the eutectic point. The eutectic point is 361°F. However, as the percentage of tin in solder is decreased and that of lead is increased, the melting point increases.

There are three grades used in electrical work: 40–60; 50–50; and 60–40. The first number is the percentage of tin, the second is the percentage of lead. The higher the tin content, the easier the flow of solder and the less time it takes for the solder to harden. It is usually easier to work with 60–40 solder than with 40–60.

How the Soldering Iron Works

Most soldering irons (but not all) work on the same basic principle as a toaster, or heater, or electric iron. A current of electricity passes through a resistive element, a special type of wire that gets very hot and which is imbedded in the body or barrel of the unit. The only connection between the tip of the iron and the body is a physical one. As the body of the iron is heated by the resistive element, heat flows into the tip.

The current flowing into the iron is in the order of 1 or 2 amperes, approximately. The connecting cord is insulated and may be covered with a fibrous braid. For some of the smaller irons, such as pencil types, the power cord is made of a flexible, rubber-like material, similar to lamp cord.

Tinning the Iron

Some accessories are needed for soldering, including scrap sections of cloth, a file, medium-grade sandpaper, and a support for the iron when it is not being used.

All irons, whether new or old, must be tinned before use. Tinning simply means putting a coat of solder on the tip of the iron. Some new irons are sold with pretinned tips with a silver finish. In that case, let the iron get hot, and as it does so, just rub the tip with scrap cloth. If the tip is not pretinned, let the iron get hot and then file the tip lightly, using a fairly fine-toothed file. This will remove most of the accumulated oxide and dirt on the surface of the tip. Rub the tip with sandpaper and finish the job by further rubbing with scrap cloth. Apply solder directly to the tip of the iron. A small globule of solder will form on its surface. Wipe it and this action will spread some of the solder along the tip face, giving it a shiny, silvery appearance. Keep repeating until all the sides of the tip are coated. Then, and only then, will soldering be possible.

The tip of the iron must have the correct temperature, neither too hot nor too cold. Some irons come equipped with automatic thermal controls. With a nonthermostatic iron, if the tip of the iron is too cold, the solder will not melt, or if it does melt, melting will occur slowly and sluggishly, forming a soft gray blob. If the iron is too hot, the solder won't adhere. The best technique is to plug the iron into a receptacle, solder the joint, and then remove the iron from its receptacle. Do not plug in an iron indefinitely while doing other electrical work. Plug the iron in shortly before soldering and remove it directly thereafter.

Soldering Copper

While solderless connectors (wire nuts) are useful for making electrical joints with minimum effort and time while providing insulation, there will be times when soldering will be necessary. An ideal joint would be one that is soldered, then reinforced with the help of a solderless connector. The combination of solder and connector gives the joint optimum mechanical and electrical security.

The watchword in soldering is cleanliness. It is difficult, if not impossible, to solder wires that carry some traces of wax, enamel, or other insulating materials.

Soldering Technique

Place the tip of the iron so that it touches the undersurface of the joint to be soldered (Figure 2-38). Keep the iron in position, then apply the solder to the top portion of the joint. It is the heated joint that must melt the solder, not the iron. Use just enough solder to cover the wire (Figure 2-39). The soldered joint should appear smooth and shiny, not lumpy and dull gray. An improperly soldered connection is called "cold" soldered.

After the solder has flowed onto the joint, remove the iron and give the solder a chance to cool. Do not move the joined wires before the solder has cooled. If you are holding the wires with one hand while soldering with the other, keep your hand motionless for a few seconds after removing the iron. If, for some reason, the soldered joint looks lumpy, reheat it and allow the excess solder to flow off onto the iron tip. Do not add more solder.

It will be necessary to tin the iron repeatedly, but this is easy to do once the first tinning is accomplished. To tin the iron again, rub the heated iron tip with some scrap cloth. It is a good idea to wipe the tip with a cloth after each soldering

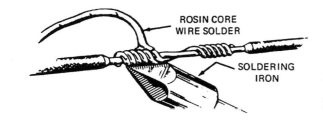

ROSIN CORE
WIRE SOLDER

SOLDERING
IRON

Figure 2-38 Correct soldering technique. Apply hot iron beneath the joint, solder above. It is the hot copper that melts the solder, not the iron.

Figure 2-39 Use the minimum amount of solder. Drawing at left shows excess solder; drawing at right indicates correct amount. The solder should fill the spaces between the wires.

job. The excess solder will help form a coating on the tip surface of the soldering iron.

The reason for tinning is to help prevent the formation of copper oxide on the tip. When an iron is heated, the hot copper tip joins with the oxygen in the air. The resulting coating of copper oxide is an insulator, preventing the adequate transfer of heat to the joint. Since the function of the iron is to make the joint hot enough to melt solder, make sure that the maximum surface area of the tip is in contact with the joint—that is, do not apply the pointed tip of the iron to the work, but rather its flat surface.

Insulating the Soldered Joint

The purpose in taping a soldered joint is to eliminate the possibility of a short between the joint and an adjacent connection or ground. Although the joint may be protected by a fuse or circuit breaker, it may be difficult to get at the joint to make a repair.

Old-style tape was made of fabric coated with an adhesive. This tape had two serious faults. The edges of the tape could unravel and the tape itself could dry out, loosening the wrap. Modern tape is made of flexible vinyl, eliminating any thread unravel, and also permitting a stretched—hence a very tight—wrap.

To insulate a soldered joint, wrap it with vinyl electrical tape. Also known as PVC tape, it is available in rolls from about 12 feet to 100 feet or more and in ½-inch and ¾-inch widths. While black is the most common finish, it is also supplied in other colors, such as red, yellow, blue, and so on. Color coding is advantageous if it is necessary to identify a particular joint.

To make a good wrap, start in about 1 inch from the exposed joint. This section will still be covered with wire insulation, and the tape will make a tight bond. Start the tape at an angle, advantageous since it will result in a natural forward movement. Do not make a rapid succession of turns. Instead, as you make each turn, pull on the tape to make the turn as tight as possible. Each lap should cover about 50 percent of the preceding turn.

Solder

Like wire, solder is available in rolls ranging in length from about 7 feet to 50 feet or more. It is also sold in the form of bars and tape. Roll solder is sometimes called wire solder. It can be identified by gauge number with gauge No. 16 as the standard size for home wiring when using an iron in the 100-watt to 150-watt class. For a pencil iron or any other low-wattage type, it is easier and better to use a thinner gauge of wire solder.

Solder Tape

Solder is available in the form of small tape strips. With these just wrap the solder tape around the joint and then heat it. The advantage of using tape is that a soldering iron is not needed. A lighted match or candle will do. However, this kind of soldering is just a temporary expedient. It is difficult to avoid burning the wire insulation, so for better work a soldering iron and 60–40 solder are preferable.

Solder Bars

Solder bars are used with extremely high wattage irons or with a torch. Such bars aren't practical for home soldering work.

Flux

When soldering copper wire, the intense heat of the iron not only helps form an oxide around the copper soldering tip but also promotes a similar oxide around the wire joint. These oxides may either make soldering more difficult or prevent it. To avoid the formation of such oxides, soldering requires the use of a flux. For home electrical work the preferred flux is rosin. Fortunately, solder is available as a rosin-cored type—that is, wire solder contains a core of rosin, so the flux is applied automatically together with the solder.

There are also various other types of fluxes, such as muriatic acid, hydrochloric acid, ammonium chloride, or zinc chloride, but these are soldering fluxes for metals other than copper wire, although acid flux can also be used on copper.

Rosin is the preferred flux for electrical work. It is easier to work with acid flux, but joints soldered with it tend to corrode.

While the rosin in rosin-core solder does keep oxides from forming on the wires to be soldered, it will not remove existing dirt or oxides. A wire must be shiny clean if the solder and the rosin flux are to be effective.

SOLDERING IRONS

Soldering irons are available in a variety of sizes, shapes and wattage ratings. For home use the most practical iron will have a rating between 100 and 150 watts. A lighter-weight iron, such as a 25- or 35-watt size is also helpful for soldering smaller gauge wires.

Irons are also available as soldering guns, irons having fixed tips, irons having replaceable tips, and battery-operated irons.

For average work around the home, a fixed-tip iron is satisfactory. An iron with replaceable tips permits doing a greater variety of soldering jobs but isn't necessary unless you expect to become a professional electrician.

Hot Knife

One advantage of the interchangeable tip feature is to permit a hot knife attachment. The hot knife consists of assorted blades that can be attached to the iron. When the blades become hot, they can be used for cutting through and removing wire insulation quite easily. This insulation is the fibrous or plastic type only and does not include enamel or similar coatings.

Battery-Operated Iron

The battery-operated soldering iron uses a single Ni-Cad (nickel cadmium) battery that can be recharged. The advantage is that you can use the iron in areas that cannot be reached by an extension cord, possibly for outdoor wiring. The disadvantages are that the battery does need recharging, is moderately heavy, and the amount of tip heat is rather small.

Soldering Gun

A soldering gun (Figure 2-40) has a gun shape—hence its name. Unlike ordinary soldering irons, the gun heats extremely quickly, and it usually has an interchangea-

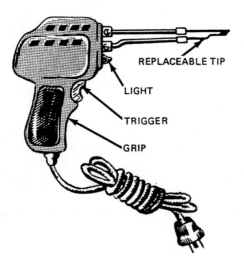

REPLACEABLE TIP

LIGHT

TRIGGER

GRIP

Figure 2-40 Soldering gun.

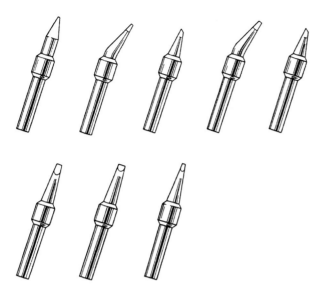

Figure 2-41 Replaceable tips for a soldering gun (Courtesy Weller).

ble tip facility (Figure 2-41). Some also have dual power ratings, such as 100 watts and 140 watts. The 140-watt rating is useful for soldering a joint consisting of a number of thick wires.

Unlike many soldering irons that must be plugged in or unplugged from the power line, the soldering gun has a trigger action. Pressing the trigger turns the gun on, releasing it turns it off. The same trigger controls either 100-watt or 140-watt heat, depending on how far back the trigger is pulled. Some soldering guns come equipped with a built-in light, helpful when it is necessary to solder in dark areas.

The tip of a soldering gun must be tinned, using exactly the same procedure as with any other soldering iron. Some soldering guns come equipped with a pre-tinned silver tip. The silver coating on the tip makes the iron ready for immediate use. Ultimately the silver wears away, then the iron must be tinned with solder.

One of the advantages of the soldering gun is that it is housed in a plastic case. This means the gun can be put down on a table when not being used. Stands are available for soldering irons and guns.

The Pencil Iron

The advantage of a pencil iron is that it is lightweight, is rated at about 25 to 35 watts, and can often be used to solder in difficult to reach areas. Pencil irons can have interchangeable tips. Some of these can be curved, straight, or extra long. (Figure 2-42).

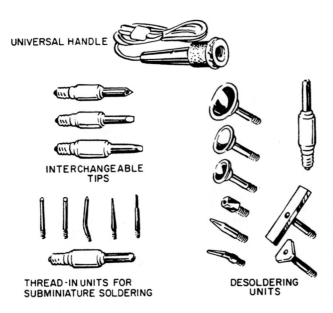

UNIVERSAL HANDLE

INTERCHANGEABLE
TIPS

THREAD-IN UNITS FOR
SUBMINIATURE SOLDERING

DESOLDERING
UNITS

Figure 2-42 Use a pencil iron for soldering fine wires. The iron can have a universal handle to accommodate interchangeable tips. For removing solder, use the desoldering units shown at the right. Typical ratings of pencil irons are from 25 to 35 watts.

Interchangeable Tips

There are two ways in which removable tips can be fastened to irons: by setscrews or by a threaded tip. The setscrew type is not as good as the screw-in type, which allows greater surface area contact between the tip and the barrel of the iron. With the setscrew type, turning the screw to make sure the tip is securely fastened results in pushing part of the tip away from the barrel, reducing heat transfer. The setscrew can also be easily lost.

The screw-in tip also has its problems. The heat can cause the threads of the tip to remain fastened to the barrel's threads. When this happens, the tip is said to be "frozen" in position, an odd term to apply to a soldering iron. To prevent this, use an "anti-seize" compound. Put some of it on the threads of the tip before turning it into the barrel of the iron.

The Soldering Iron Stand

When not soldering, you will need to rest the iron on some surface. Even when the plug is removed or the trigger is released on a gun, the iron will remain hot for some time, creating a potential source of trouble. Accidentally touching the exposed metal barrel of the iron can produce a painful burn. If the iron is put on a combustible surface, such as wood, it can result in a fire or make an ugly burn mark.

It is always best to use a soldering-iron stand. You can make a temporary stand by crushing a V notch into any clean tin can, but a stand made for the purpose is better. Such a stand will completely enclose the metal portion of the iron. Made of metal in grille shape, it prevents the iron from touching the table and also acts as a barrier between the hands and the hot metal. The open grille allows the heat of the iron to escape.

Cooling the Iron

Do not cool the iron by wetting it or by completely immersing it in water. The best way to get rid of the heat in a hurry is to rest the metal shank or barrel of the iron on as large a metal surface as possible. That surface should be clean and free of paint or dirt.

The Solder Sucker

It takes experience to be able to use exactly the right amount of solder. With excess solder a large blob forms and the result is a cold-solder joint. When this cools, it is difficult to remove the excess. One way of removing excess is to reheat the joint and try to brush away some of the unwanted solder with the hot tip of the iron. There are several risks. Some of the solder may land on the wire's insulation, burning it. Another possibility is that it may drop on the table, the floor, or even worse, on some part of the body. If it lands on clothing, it can be very difficult to remove.

A better technique is to use a solder sucker (Figure 2-43), which consists of a squeeze bulb and a tip. Its function is to suck up excess solder. As it does, it also removes some of the heat from the joint being soldered. Do not use glass or plastic medicine droppers as temporary substitutes.

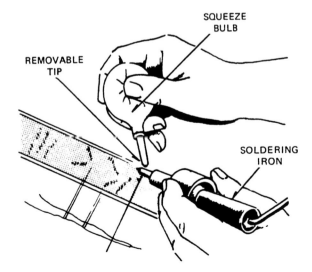

Figure 2-43 Solder sucker.

Soldering-Iron Hints

It isn't difficult to use a soldering iron. Soldering is easy by following some simple rules.

1. Do not try to use a cold iron or one that hasn't reached its correct operating temperature. If the iron isn't hot enough, the solder will not melt into its liquid state but will become a gray mass. This blob may actually separate the wires that are to be joined, and the result may be an open circuit. It may be difficult to locate the problem, since the blob will prevent inspection of the joint.
2. The soldering-iron tip must be tinned. If at all possible, tin the tip with silver— that is, hard solder it. If not, tin it with rosin-core soft solder. It is better, although not easier, to use rosin core rather than working with solder with an acid flux.
3. During use, the tinning on the tip of the iron wears away, and it will be necessary to tin the tip again.
4. While the iron is hot, wipe the tip regularly with a cloth. The cloth should be clean and preferably have a rough texture. Do not work with a cloth that has been previously used to wipe up oil spills or paint.
5. Cleanliness is the watchword in soldering. The copper wire to be joined must be clean. The soldering-iron tip must not only be tinned, but must also be clean. Grease, oil, dirt, oxidation, bits of insulating material, and enamel can and will prevent soldering.
6. When not in use, rest the iron on a stand.
7. Do not try to gauge the temperature of the iron by bringing it close to your face.
8. Do not leave the iron plugged into an outlet for an indefinite period of time. Poor soldering can result if the iron is too hot as well as too cold. This does not apply to soldering guns, which turn off automatically when the trigger is released.
9. Shaking the iron to remove excess solder is an extremely dangerous procedure.

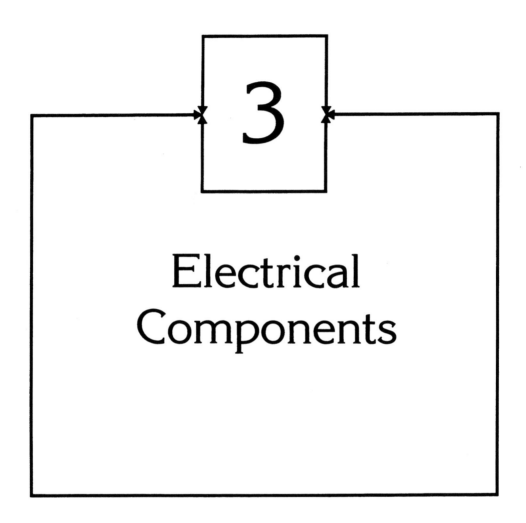

3

Electrical
Components

An electrical component is any part of a wiring system. There are two types: those that are integral to the system and those which are add-ons or accessories. The service box, the branch wiring, receptacles, and electric boxes are integral; that is, they form part of the system, and without them the system would not be complete. An add-on is any component that is used to augment the system. A night light is an add-on; the system will work with or without it. An extension cord is an add-on, and so is a cube tap, or a soldering iron.

In some cases it is difficult to differentiate between an integral unit or an add-on. Electrically operated devices can be plugged in or not, so in this sense they are separate from the wiring system even though they may be connected permanently. They do not add to the functioning of the system, but they do make use of it. Fuses are integral, since the system will not work without them, even though they are replaceable items.

A ground-fault circuit interruptor (GFCI) is plugged in permanently, becomes part of the house circuitry, and yet is an add-on since the system will work with or without it. Each component, then, must be considered individually.

ELECTRICAL SYMBOLS

Many of the electrical jobs around the house are quite simple. Replacing a switch or a receptacle and installing a dimmer or replacing a fuse are easy.

For extensive wiring changes it may be helpful to draw some sort of wiring plan. This plan can consist of pictorial drawings, actual pictures, of the wiring and the parts, such as light bulbs, switches, fuses, outlets and fixtures. An easier method, and one that will be more accurate, is to use electrical symbols.

An electrical symbol is a kind of shorthand. A knowledge of these symbols will be helpful to those who plan to continue into an electrical apprenticeship on the way to becoming electrical technicians. This knowledge is also useful to home owners who can manage to get a copy of their home electrical wiring plans. Such a plan is helpful when considering an expansion of an existing electrical system or even for making minor changes. It will also be useful when trying to solve in-home electrical problems involving the wiring system.

Figure 3-1 shows the commonly used symbols. Wiring drawings will sometimes consist of a combination of pictorials and symbols.

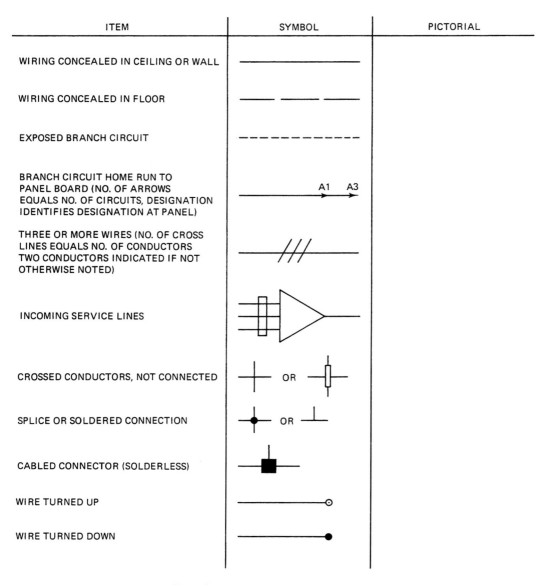

ITEM	SYMBOL	PICTORIAL
WIRING CONCEALED IN CEILING OR WALL		
WIRING CONCEALED IN FLOOR		
EXPOSED BRANCH CIRCUIT		
BRANCH CIRCUIT HOME RUN TO PANEL BOARD (NO. OF ARROWS EQUALS NO. OF CIRCUITS, DESIGNATION IDENTIFIES DESIGNATION AT PANEL)	A1 A3	
THREE OR MORE WIRES (NO. OF CROSS LINES EQUALS NO. OF CONDUCTORS TWO CONDUCTORS INDICATED IF NOT OTHERWISE NOTED)		
INCOMING SERVICE LINES		
CROSSED CONDUCTORS, NOT CONNECTED	OR	
SPLICE OR SOLDERED CONNECTION	OR	
CABLED CONNECTOR (SOLDERLESS)		
WIRE TURNED UP		
WIRE TURNED DOWN		

Figure 3-1   Electrical parts, symbols, and pictorials.

ITEM	SYMBOL	PICTORIAL
SWITCHES: SINGLE-POLE SWITCH	S	
DOUBLE-POLE SWITCH	S_2	
THREE-WAY SWITCH	S_3	
SWITCH AND PILOT LAMP	S_P	
CEILING PULL SWITCH	Ⓢ	
PANEL BOARDS AND RELATED EQUIPMENT PANEL BOARD AND CABINET		
SWITCHBOARD, CONTROL STATION, OR SUBSTATION		
SERVICE SWITCH OR CIRCUIT BREAKER	■ OR ■ OR ⊗	
EXTERNALLY-OPERATED DISCONNECT SWITCH		
MOTOR CONTROLLER	OR MC	
MISCELLANEOUS: TELEPHONE		
THERMOSTAT	Ⓣ	
MOTOR	Ⓜ	

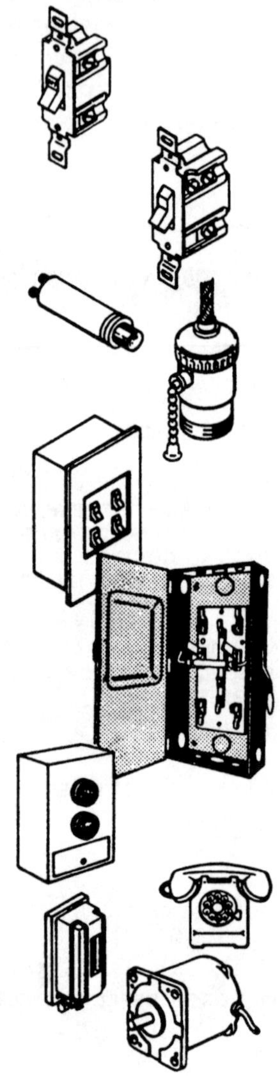

Figure 3-1 (continued)

ITEM	SYMBOL	PICTORIAL
LIGHTING OUTLETS*: CEILING		
WALL		
FLUORESCENT FIXTURE		
CONTINUOUS ROW FLUORESCENT FIXTURE		
BARE LAMP FLUORESCENT STRIP		
RECEPTACLES**: SINGLE		
DUPLEX		
QUADRUPLEX		
SPECIAL PURPOSE		
20-AMP, 240-VOLT		

SINGLE FLOOR RECEPTACLE
  (BOX AROUND ANY OF ABOVE
  INDICATES FLOOR RECEPTACLE
  OF SAME TYPE)		

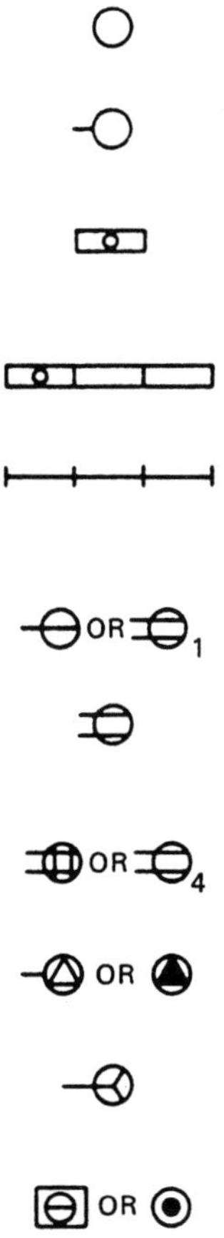

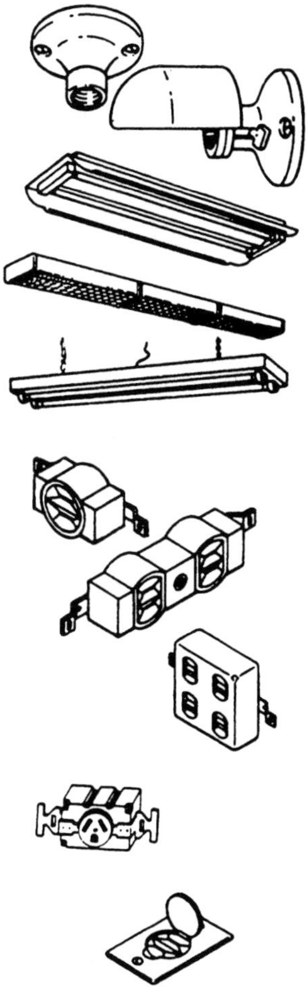

*LETTERS ADDED TO SYMBOLS INDICATE SPECIAL TYPE OR USAGE
J – JUNCTION BOX
L – LOW VOLTAGE
R – RECESSED
X – EXIT LIGHT
**LETTER G NEXT TO SYMBOL INDICATES GROUNDING TYPE

Figure 3-1   (continued)

PLUGS

Plugs, sometimes known as male connectors or simply connectors, are used as a link between a receptacle and connecting wires. Many home appliances and extension cords are terminated in plugs. These are inserted into receptacles enabling electrical power, available at the receptacle, to be delivered to some appliance.

Plug Types

Plugs can be permanent types or removable. Permanent plugs are molded onto connecting wires; removable plugs can be joined to or removed from wires.

The Basic Two-Prong Plug

This type is the simplest, probably the most widely used but least desirable of all the plug types. It can be supplied molded onto a line cord, or it may be a separate item, known as an attachment plug, designed for mounting on a two-wire cord.

As shown in Figure 3-2, the plug has a pair of equally dimensioned prongs (sometimes called poles), so it can fit in either of two ways into an outlet. It is nonpolarized and does not make use of a grounding terminal.

Two-prong plugs that are molded directly onto a line cord may not have enough strength to keep the prongs firmly in position, especially if the plug is made of a rubber-like material and is subject to repeated use. Plugs using a plastic surround are better, since they do not yield as easily to pressure. Consequently, depending on how it is made, the plug may make poor, intermittent contact with its corresponding outlet. Spreading the prongs is often not helpful. Instead, put a small crimp into one of the prongs, using gas pliers.

Some of these plugs do not have enough body to permit easy removal from

TWO-PRONG
PLUG

Figure 3-2   Two-prong, nonpolarized molded plug.

an outlet. Pulling on the line cord can damage the connection between the wires of the cord and, in a worst-case situation, can produce a short. If the plug is trouble-some, replace it with one having a molded rubber grip (Figure 3-3), or a plastic type that permits a good finger grip.

Figure 3-3 Two-prong, nonpolarized rubber molded plug with finger grip.

Connecting Lamp Cord to a Two-Terminal, Nonpolarized Plug

The purpose of making connections to a plug is electrical only. It is true that wrap-ping wires around the screws of a plug is a mechanical action but at no time should there be a strain on the connecting wires. To relieve the possibility of mechanical strain, it is a good technique to tie a knot in the connecting wires, as indicated in (a) in Figure 3-4. The recommended knot is known as an Underwriters' knot. Prior

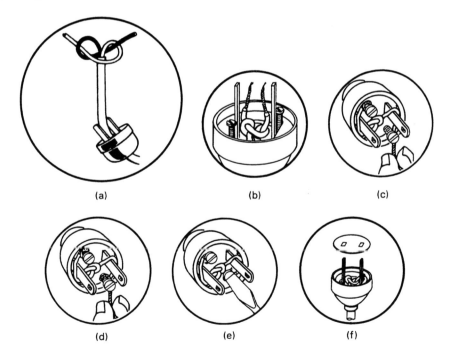

(a)   (b)   (c)

(d)   (e)   (f)

Figure 3-4 Connecting lamp cord to a two-prong, nonpolarized plug.

to making the knot, push the lamp cord through the opening in the plug, then strip
the two wire ends. After making the knot, pull on the cord until the knot is in the
space between the two prongs, as in (b). (c) to (f) show the remainder of the steps.
Each wire is pulled around one of the prongs and is then fastened by a screw. The
wrap around the screw, a captive type, is in a clockwise direction.

The Round-Cord Plug

This two-prong plug is equipped with a metal extension and is intended for use with
round cords. The extension consists of a pair of brackets held together by screws at
their ends, as shown in Figure 3-5. The brackets can be used for the easy removal
of the plug from its outlet and also work as a strain relief. The plug itself has a
round shape but this can be an interference when the plug is to be connected into a
multiple receptacle.

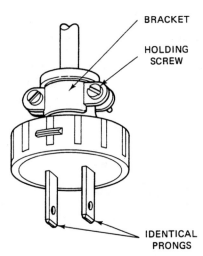

BRACKET

HOLDING
SCREW

IDENTICAL   **Figure 3-5** Bracket connected,
PRONGS   nonpolarized plug for round cable.

The Polarized Plug

A polarized plug (Figure 3-6) is one in which one of the two prongs is wider than
the other. As a result it will only fit into a receptacle that is also polarized. The
design of the prongs is such that they exert a spring like action, ensuring better
electrical contact.

   The wider of the two prongs is the neutral and ensures connection to the neutral wire leading to a switch. The corresponding receptacle has one hot slot, which
is the narrower of the two. The other, larger slot, is the wider one. Using the polarized plug ensures that the black color-coded wire will be the one that is opened or
closed by the switch. The neutral wire is the one that must always be uninterrupted.

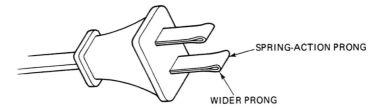

Figure 3-6 Polarized, two-prong plug.

The Quick-Connect Plug

A quick-connect plug simply requires the insertion of lamp cord and the closing of a snap feature to complete the connection. A pair of sharp-pointed connectors inside the plug pierce the insulation of the cord, contacting the wires of the cable. No wire stripping is required.

Quick-connect plugs are a two-prong type and can be polarized or nonpolarized. They can also be on-axis or off-axis types. With the on-axis arrangement the connecting lamp cord feeds directly into the plug. Figure 3-7 shows an off-axis type and is one in which the lamp cord is at right angles to the plug. This setup is desirable when the cord is to be fastened to a baseboard.

The illustration shows the steps to follow in connecting the cord to the plug: (a) is the plug and (b) is the cord. Note that the cord is ribbed, a way of identifying

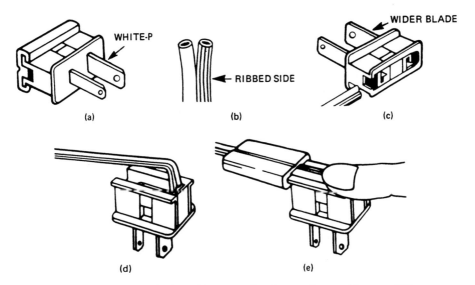

Figure 3-7 Right-angle plug with automatic cable attachment. (*Courtesy Gilbert Manufacturing Co., Inc.*)

the hot wire and the neutral. (c) shows how the wire is inserted into the plug. The neutral wire is at the top; the hot lead is below it. In this way the neutral wire will be connected to the wider blade. Following the wire insertion, it is bent down to fit into a channel at the top of the plug as in (d). A sliding tab (e) is made to slide along this channel and to force the two wires to make contact with the prongs. Sharp bits of metal cut through the wire insulation to make contact with the conductors.

Three-Prong Plugs

This type of plug, and there is a considerable variety, is intended for three-wire cables consisting of a hot wire, a neutral, and a ground wire. These plugs are rated for voltage- and current-handling capability.

The plug consists of three prongs, two of which may or may not have equal dimensions. The third prong has a tubular shape and is the ground connection. If the plug is held as in Figure 3-8, the prong at the right is the neutral while that at the left is the hot connection. The tubular prong is shown at the top in this drawing. The receptacle, however, can be mounted so that the tubular entrance is at the bottom. The positioning of the receptacle does not alter the polarity of any of the prongs.

THREE-PRONG PLUG   **Figure 3-8** The three-prong plug.

Two types are shown in Figure 3-9. One is a 15-ampere plug; the other is a 20-ampere type, with both rated for 120-volt input. The 15-ampere plug has a pair of prongs that are parallel to each other, while the 20-ampere plug has the same prongs mounted at right angles. The single tubular extension used on both plugs is the ground connection. Since the plug can be inserted in just one way into its receptacle,

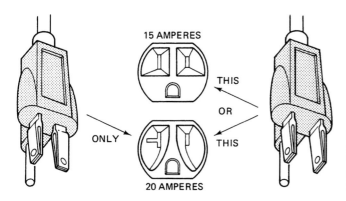

Figure 3-9 15-ampere (right) and 20-ampere (left) receptacles and plugs. The 15-ampere plug can fit into either receptacle. The 20-ampere plug can fit into a 20-ampere receptacle only.

the active prongs are the same size. The circular symbols in Figure 3-10 represent the plug and its receptacle. The letter R is an abbreviation for receptacle; P is for plug. The 15-ampere plug is capable of fitting into either the 15-ampere or the 20-ampere receptacle. The 20-ampere plug, however, can only fit into a 20-ampere receptacle. The design of the prongs supplies information about the current capacity of the power branch to which they are connected.

Figure 3-10 Symbols for 15- and 20-ampere receptacles (R) and plugs (P). W is white; G is ground. The remaining slot or prong (B) is the hot lead.

Higher-Current, Double-Voltage Plugs

Figure 3-11 illustrates still another three-prong plug, but this one is current rated at 30 amperes. Another feature is that it has a two-voltage capability, supplying either 120 volts or 240. The higher voltage consists of two 120-voltage sources in series.

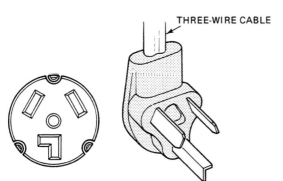

THREE-WIRE CABLE

Figure 3-11 Three-prong plug rated at 120 or 240 volts, 30 amperes. Receptacle is shown at the left.

The plug is designed for three-wire cable: a black lead carrying 120 volts, a red lead for 240 volts, and a neutral wire. Each hot-lead prong is mounted at an angle, while the neutral prong is made of two lengths of metal at right angles.

The electrical symbols are shown in Figure 3-12. The letter W represents the neutral, with X and Y the two hot leads. R is the symbol for the receptacle, and P is that for the plug. The two hot-lead prongs have the same shape and thickness.

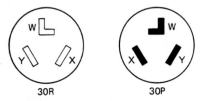

Figure 3-12 Electrical symbols for high current, double-voltage plug. W is neutral; X and Y are hot leads.

High-Voltage, High-Current Plug

Unlike the preceding plug, this one (Figure 3-13) is intended for 240 volts only. It has a pair of flat, similarly shaped prongs, one of which is the hot lead, the other the neutral. The metallic ground prong has a semicircular shape.

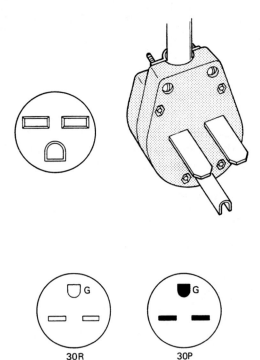

Figure 3-13 High voltage (240 volts) and high current (30 amperes) receptacle (left) and plug (right). The symbols for the receptacles (R) and the plug (P) are shown below.

50 Ampere Plug

Figure 3-14 is that of a 50-ampere plug and a head-on view of its receptacle at its left. This plug is a dual voltage type and can use 120 and 240 volt connections on its receptacle. Its two hot terminals are mounted at angles to each other, while neutral is a single, vertically positioned prong. The electrical symbols for the receptacle and the plug are shown in drawing (b). Drawing (c) is a pictorial of the receptacle.

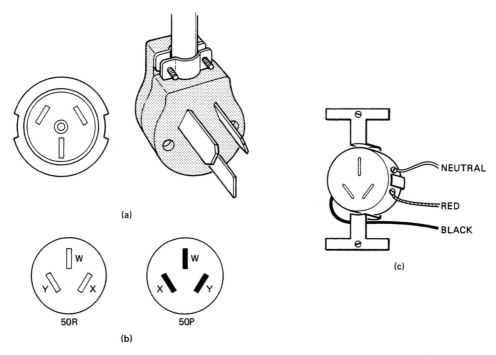

Figure 3-14 50-ampere receptacle (a), top left, and its plug (a), right. Symbols for the receptacle (R) and its plug (P) shown in (b). W is the neutral; X and Y are the hot leads (black and red). Connections shown in (c).

Adapter Plugs

Some homes are not equipped to accommodate three-prong plugs but will accept two-prong nonpolarized types only. A three-prong plug can fit into a two-prong receptacle by using an adapter, as in Figure 3-15. The adapter has either a wire coming out of it or a lug. Either of these is a ground connection, and its function is to ground the connected electrical device. Connect the wire or the lug to the center screw that holds the receptacle cover in place.

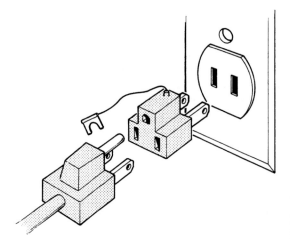

Figure 3-15 Adapter to permit three-prong plugs to use two-slot receptacle.

An adapter provides a temporary three-hole outlet. The trouble is that once an adapter is used it may assume permanent status. While an adapter is better than no adapter at all, the best arrangement is to replace two-hole receptacles with three-hole types.

RECEPTACLES

A receptacle is an electrical component designed to accommodate a plug for the delivery of power to some electrical component. A receptacle can be mounted in a wall, in which case it is positioned inside an electrical box made of metal or plastic, or it can be designed for surface wall mounting. For the latter use, it is completely enclosed in a plastic container. (The words receptacle and outlet are used interchangeably.)

The Two-Prong Receptacle

At one time this was the most commonly used receptacle (Figure 3-16), but it is gradually being replaced by a three-prong type. The two-prong receptacle accepts the insertion of a two-prong plug.

A receptacle is a terminal for power wires leading back to the fuse or circuit breaker box and can be connected anywhere along that line. It is wired in parallel with its power line and, for this kind of receptacle, consists of a hot or black lead and a white or neutral lead. While single receptacles are available, most often they are duplex, consisting of one receptacle mounted directly above the other.

These older receptacles are screw types only; that is, the two wires of the power

Figure 3-16 Receptacle for nonpolarized plug.

line are connected by screws. More modern receptacles give the user an option of either screw or push-in connections.

Anatomy of a Receptacle

Details of the front structure of a duplex receptacle are shown in Figure 3-17. The unit has two pairs of wings, one across the bottom, the other across the top. These fit against the plasterboard or dry wall and help keep the receptacle firmly in position.

This receptacle is supplied with two pairs of screws for wire connections, but some receptacles omit them and use push-in connections exclusively. Usually a receptacle has both screws and push-ins, giving the user an option. The screws on the right that are brass colored are for connections to the black or hot leads of the branch power wires leading from the fuse or circuit breaker box. The two screws on the left that have a nickel color are for connections to the neutral wire. The remaining screw positioned near the bottom left of the receptacle may be colored green. This is the ground connection. Sometimes a clip is used instead with the bared

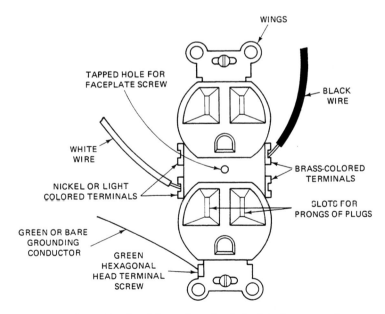

Figure 3-17 Details of front view of polarized duplex receptacle.

ground wire slipped in and under it, held in place by spring action against the metal box holding the receptacle.

Whether using screw or push-in terminals, be careful when wiring to the receptacle terminals. Transposing the black and white wire connections will result in a short circuit and will blow the fuse or open the circuit breaker. It could also damage the receptacle, requiring its replacement. This precaution is applicable to all receptacle types.

The screws are captive types and can be opened just enough to accommodate connecting wires.

The receptacle has four slots, two above and two below. These are for the insertion of the prongs of a polarized plug. The slot at the upper right is smaller than that at the left. The two lower slots have the same construction. The slots at the right are hot and are internally connected to the brass-colored terminals.

The receptacle has a threaded hole in its center. This accommodates a small screw for holding the wall plate in position. This screw has the same color as the wall plate, and it will be troublesome to replace if dropped and lost.

Not shown in the drawing are various bits of data, including the UL logo, voltage, and current ratings.

Rear of the Receptacle

Like the front, the rear of the receptacle may supply useful information, but the amount will vary, depending on the manufacturer. Useful data may include information on the kind of wire to use such as "No. 12 or No. 14 solid copper wire only."

Some receptacles are equipped with a strip gauge. This is an aid in determining the amount of insulation to be removed from a wire to be used as a push-in connection. It has a double purpose: to make sure the stripped wire is long enough to make good contact with the connecting elements inside the receptacle and to avoid any unnecessary exposure of bare copper.

For a duplex outlet there will be four push-in connections—two for each outlet. There will be a greater number depending on the number of outlets used in the receptacle assembly.

The receptacle will be equipped with a pair of holes or rectangular slots for the release of wires that are push-in connected. If holes are used, insert a short length of copper wire to operate the release. For rectangular slots use a small screwdriver. Hold the wire or screwdriver in place and then tug on the connected wire.

Recessed Receptacles

After inserting a plug into a receptacle, the connecting wires of the plug extend away from the wall holding the receptacle. For many applications, such as lamps or toasters, this arrangement is satisfactory. There is no reason why the wire should hug the wall. But there are times when it is desirable to have a wire as close against a wall as possible, as in the case of wall-mounted clocks. Special clock receptacles are available. These units are recessed and will hold both the plug and excess wire.

Twist-Lock Receptacles

A loose plug that repeatedly falls out of its receptacle is a nuisance, especially if the receptacle isn't conveniently located. It may be positioned behind a bed and below the bed frame. Spreading the prongs of the connecting plug may help, if the plug is made of plastic which does not yield to pressure. A rubber plug may defeat the prong-spreading action.

For worst-case situations use a twist-lock plug and an associated twist-lock receptacle, as shown in Figure 3-18. Such plugs and receptacles require a twist-and-turn push action, so the plug not ony remains secure in the outlet but also makes more positive contact.

Figure 3-18   Twist-lock receptacles.

Receptacle Power Ratings

Receptacles intended for in-home use have power-handling ratings, but this may be expressed in terms of current instead of wattage. To determine its power capability, multiply the amount of line voltage by the maximum rated current of the receptacle. For a 15-ampere line the power is $15 \times 120 = 1800$ watts or 1.8 kilowatts. This is for the entire receptacle, and if it is a duplex type, then each section will have a rating of $^{1800}/_2$ or 900 watts.

This does not present a problem if the outlet is used by a single 75-watt floor lamp or an equally rated radio set or even a 300-watt TV. But it is very easy to convert the duplex receptacle into a quadruplex by using taps. With so many outlets available with the help of the tap, and by plugging in more and more appliances, it is easy to exceed the power rating of the receptacle.

Back Wired vs Front Wired

A receptacle can be described in a number of ways: single, duplex, triplex, and so on; its current or power rating; whether grounded or not grounded; or the number and type of slots. In some instances it is referred to as back wired or front wired.

A back-wired receptacle is a push-in type with all connections (except ground) made at the back; a front wired is one that uses screw connections, but these are actually at the side. Aside from the use of plugs and adapters, receptacles do not have front connections.

The Switched Receptacle

Instead of a duplex receptacle, it is possible to use a combined switch and receptacle with the switch replacing one of the receptacles in the duplex. The wiring can be arranged so that the switch controls power to the outlet. With this type of receptacle the switch must be in its on position for power to reach the appliance plugged into the outlet.

This unit is known as a switched receptacle. All other types are called un-switched.

Sockets vs Receptacles

Although regarded as a separate category, electrical sockets, although different in shape, are considered part of the receptacle family. The difference is that sockets are screw-in or bayonet types, while receptacles depend on sliding contact between the prongs of a plug and the metal arms of a receptacle. Because they are so closely related, sockets are sometimes called outlets or receptacles.

Safety Inserts

Most often receptacles are mounted very low on a wall near its baseboard. As a result the receptacle is within reach of a child's busy, inquisitive fingers. While receptacles are designed not to permit the insertion of fingers, children can push in spoon handles, scissors, or other metal objects.

To prevent this, use plastic inserts. These inexpensive items fit into the slots of an unused receptacle and cover them completely. The fit should be tight enough that the inserts are not easily removed. The inserts are also useful in winter and help keep cold air from findings its way into the house or apartment through a receptacle and its slots.

Some, known as safety receptacles, are designed to prevent the accidental insertion of any device except a plug.

Receptacle Dimensions

Figure 3-19 shows the mounting dimensions for a three-wire straight-blade grounding type receptacle that is side wired. The center hole is for fastening a wall plate to

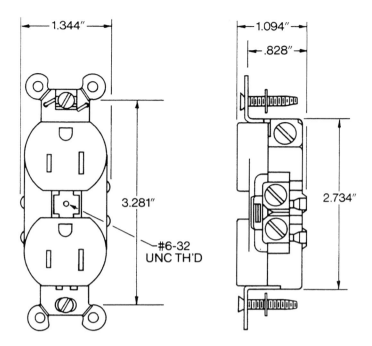

Figure 3-19 Front and side views of receptacle showing dimensions. (*Courtesy Hubbell Incorporated, Wiring Device Division*).

the receptacle and uses a #6-32 machine screw. The illustration shows a front view (left) and a side view (right).

Surge Suppression Receptacles

Personal computers and word processors are at the mercy of transient voltage surges. These surges can result in permanent loss of data or can do even more damage, to the point of incapacitating computers, word processors, and disk drives.

Many surge suppressors are external devices that plug into standard AC outlets. Another type, a more permanent arrangement having a larger margin of safety, is the surge receptacle.

A surge suppression receptacle is available as a "single boss" type, accommodating a single plug, or it can be a "two-boss" unit. The double type, Figure 3-20, is equipped with a power-on indicator light that instantly verifies that power is available at the receptacle and that the suppression circuit is fully functional. If the light goes off, power has been interrupted or the surge protection section has been damaged.

The double receptacle unit has a damage alert beeper that beeps when the surge protection is no longer functioning and keeps beeping until the receptacle is replaced. The component not only absorbs and dissipates transient surges but also

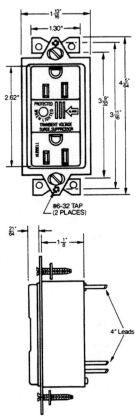

Figure 3-20 Duplex surge-suppression receptacle. (*Courtesy Hubbell Incorporated, Wiring Device Division*).

provides a measure of radio-frequency interference (RFI) protection. It fits a standard in-wall electric box and easily retrofits a standard receptacle. The wall plate can be available in a strong, distinctive color, such as deep blue, to distinguish it from ordinary plates, making the surge suppression receptacle immediately recognizable.

WALL PLATES

A wall plate supplies a finish, a covering for switches and receptacles, and is made to accommodate any one of the large variety of these components. At one time little attention was paid to these plates, but now they are available in numerous colors and designs to fit in with home decor.

Some of the colors and finishes are bank bronze, statuary bronze, satin bronze, polished brass, satin brass, clear anodized aluminum, black anodized aluminum, dark bronze anodized aluminum, satin aluminum, satin stainless steel, polished stainless steel, painted metal in red or black, wrinkle painted metal in white or brown or ivory, antique copper and gold, and hammered copper. Some plates are made of Lexan thermoplastic, and these are available in red, gray, ivory, brown and white.

Wall plates are made for switches ranging from single to as many as ten, for single or multiple receptacles, and for combined switches and receptacles. Some are various sizes of rectangular blanks to cover the space formerly used by a switch or receptacle (Figure 3-21).

Figure 3-21 A few of the numerous types of electric box wall covers. (*Courtesy Mulberry Metal Products, Inc.*)

Steel wall plates are not only available in a variety of colors, but in various finishes as well, including semi-gloss, matte, smooth, hard baked wrinkle, mirror and dull.

Switch wall plates are also available in a large number of sizes, referred to as standard, maxi and jumbo. Figure 3-22 shows some representative dimensions. The distance of the mounting holes can also vary. To make certain the size is correct

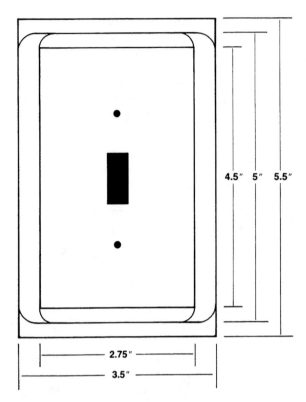

Figure 3-22   Dimensions for switch wall
plates. (*Courtesy Mulberry Metal
Products, Inc.*)

(prior to making a purchase of a new plate) take along the old wall plate. If this
isn't possible, make a simple sketch indicating the length and width dimensions and
the distance between the mounting holes. Alternatively, take the switch along as a
guide when purchasing a wall plate. Typical separations of the mounting screws are
3.281 inches, 2.375 inches, and 3.812 inches, as indicated in Figure 3-23. For multiple switches, consult the illustration in Figure 3-24.

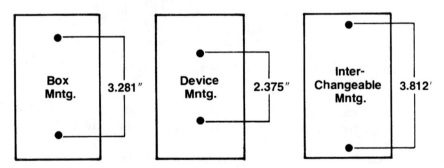

Figure 3-23   Screw hole mounting dimensions for wall plates. (*Courtesy
Mulberry Metal Products, Inc.*)

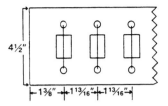

Figure 3-24   Wall plate for multiple switches.

Maxi Wall Plates

These can be used with plastic electric boxes, oversized boxes to accommodate Romex, or where the wall has irregularly cut holes or broken plaster.

Jumbo Oversize Plates

Use these for covering broken plaster and irregularly cut holes in sheet rock or paneling around the electrical box. They can be used to protect the wall surrounding wiring devices, thus reducing maintenance. They are also useful for the replacement of wall plates in old work where a mark is left around the original plate.

Deep Plates

These are ideal for use when the electric box extends out from the wall. The distance can be $\frac{9}{16}$ inches or $\frac{3}{8}$ inches as indicated in Figure 3-25.

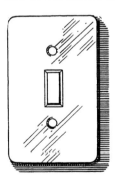

Figure 3-25   Deep plates. For use where electric box extends from wall. Left $\frac{9}{16}$ inch; right $\frac{3}{8}$ inch. These are available for one, two or three toggle switches; for single and duplex receptacles; and for blanks. (*Courtesy Mulberry Metal Products, Inc.*)

Narrow Plates

Use these for movable office partitions, trailers, mobile homes, and marine installations. These plates have heights of 4-$\frac{1}{2}$ inches, 6-$\frac{1}{8}$ inches, and 8-$\frac{1}{8}$ inches. See Figure 3-26.

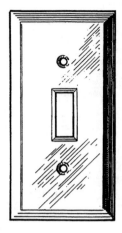

Figure 3-26 Narrow wall plate. Available heights are 4-½ inches; 6-⅛ inches; 8-⅛ inches. (*Courtesy Mulberry Metal Products, Inc.*)

Plates for Oil and Gas Burners

Switches placed in electric boxes can be mounted on conduit containing electric cables leading to oil or gas burners. The cover plates for these units may carry useful information as shown in Figure 3-27. Cover plates are also made for air-conditioner switches and for emergency purposes. An emergency plate may also be covered with a switch guard, as in Figure 3-28, to prevent accidental use of the switch.

Figure 3-27 Oil burner (left) and gas burner (right) emergency switch wall plates. (*Courtesy Mulberry Metal Products, Inc.*)

Figure 3-28 Emergency switch guard. (*Courtesy Mulberry Metal Products, Inc.*)

Wall plates are available which indicate that the receptacle or switch is to be used for emergency purposes. These are intended for single or duplex switches, as indicated in Figure 3-29, or for duplex or quadruplex receptacles. They are available in white or black. Since most wall plates do not use black, this is a distinctive color emphasizing emergency use.

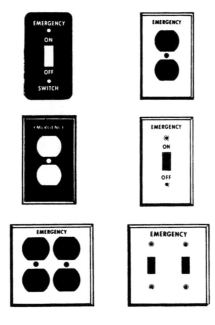

Figure 3-29 Emergency switch and receptacle wall plates. (*Courtesy Mulberry Metal Products, Inc.*)

Insulating Gaskets

These can be used beneath the wall plate to minimize the flow of cold air from the electric box beneath the plate. They are made for both switches and receptacles (Figure 3-30). They are made of close-cell polyethylene insulating foam.

Figure 3-30 Energy-saving insulating gaskets. (*Courtesy Mulberry Metal Products, Inc.*)

Mounting the Wall Plate

Wall plates for single switches are mounted through the use of a pair of screws, one positioned directly above the switch; the other below it. Plates for duplex outlets are held in place by a single, centrally positioned screw.

ELECTRIC BOXES

The basic function of an electric box is to work as an enclosure for electrical connections, but it is also used to house switches and receptacles. Boxes can be made of plastic or metal and can be mounted in wall or on the wall surface. Those intended for external use must be waterproof. In all instances the boxes must be supplied with a cover.

Naming the Box

Electric boxes can be identified in a number of ways: by size, shape, dimensions, use, location, and material of which the box is made.

Metal vs Plastic Boxes

Both types have advantages. Metal boxes can be grounded, and if metallic conduit or metal-sheathed cable is used, grounding becomes automatic. Grounded metal boxes are automatically safe and are electrically neutral. They also supply a certain amount of protection against fire.

Plastic boxes are electrically neutral, since they are nonconductive, hence do not require a grounding connection. Unlike metal boxes they are rustproof but can be more easily damaged. Like metal boxes, they require a cover.

Junction Boxes

The purpose of a junction box is to supply a housing for wires that are to be joined, whether by a splice or by soldering. The fact that a wire nut is used does not eliminate the need for a junction box. This is applicable even if the wire joint uses a wire nut covered with electrical tape.

The shape of the junction box is not significant and can be rectangular, octagonal or any other style.

GROUND

The white or neutral wire in an electrical system is connected to ground, but if for some reason this ground connection should be removed from the neutral wire, you would then have two hot leads. Touching either wire while touching ground would result in an electrical shock.

To make sure that the white wire is always grounded (that is, at zero voltage), cables now come equipped with a third wire which functions solely as a ground wire. Look on it as a backup for the white or neutral wire. Consequently, when wiring a metal electrical box, make sure to connect the bare ground wire to some metal portion of the box.

Consider the ground wire as an added safety precaution. The ground wire is especially important for nonmetallic cable, such as plastic-sheathed types.

Boxes are made of steel or plastic. Plastic boxes sell for less than steel but have the disadvantage that plastic is nonconductive so does not supply convenient grounding points. In the event of a fire, the plastic will either deform severely or may drop away, exposing wire to the flames. Steel boxes offer far more fire protection.

Plastic boxes can be purchased in ganged form, but they cannot be disassembled to make a ganged box as can be done with the metal type. NM cable is the only kind that can be used with plastic boxes. Some local electric codes prohibit the installation of such boxes.

ELECTRIC BOX TYPES

Electric boxes are available in different sizes and shapes. Boxes are named not only for these two physical characteristics but also for their ultimate use.

Octagon Boxes

Octagon boxes are intended for fixtures or for use as junction boxes. These boxes can be made of steel and are equipped with knockouts, generally ½ inch to ¾ inch in diameter and may be supplied with cable clamps. The cable clamps are somewhat similar to straps, such as those used for conduit, but are built into the electric box and can be used to hold one or a pair of cables firmly in position simply by turning a machine screw.

The octagon box (Figure 3-31) can be covered with a plate that is also octagonal in shape. The box has a pair of wings which have tapped holes for accepting a pair of screws for holding the cover plate in position. Because of their shape, octagon boxes aren't suitable for wall fixtures such as switches and receptacles.

Figure 3-31 Octagon box. This type is mostly used as a junction or a fixture box.

Knockouts

A knockout (Figure 3-32) about ½ inch to ¾ inch in diameter has a circular shape and is almost completely cut around the full length of its circumference. It can be pried open with a screwdriver to allow the entry or exit of power cables. If not removed, it offers some protection for the wires enclosed in the box.

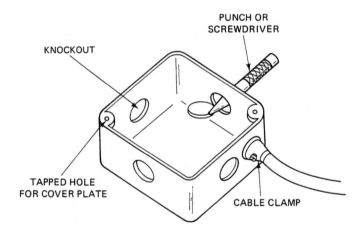

Figure 3-32 Knockouts are used to permit passage of input or output power cables.

Rectangular Boxes

So called because of their shape, these are especially intended for wall switches and receptacles (Figure 3-33). Some are designed to permit ganging by removing a side plate from each of the boxes. This doubles the available volume and supplies more room for cable occupancy.

The box has a pair of tapped holes, front top and bottom center, to accommo-

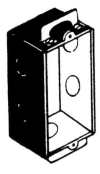

Figure 3-33 Rectangular electric box is primarily used for switches and receptacles. It can also be used as a junction box but does not provide as much interior space as an octagon box.

date screws holding receptacles or switches in position. Rectangular boxes are intended for in-wall positioning. These boxes can also be used as junction boxes, and in that case they are supplied with a steel cover plate which can be screwed into position using the tapped holes. Their disadvantage is that they do not have as much volume as octagon boxes.

Beveled Corner Box

One of the problems in doing electrical work in older buildings is the extensive use of laths, long wooden strips used as wall and plaster supports. Newer buildings are constructed of dry wall nailed against wooden studs or joists. Beveled boxes (Figure 3-34) look like rectangular boxes except that their rear section is beveled. These boxes can be replaced with rectangular types if the beveled box isn't available.

Figure 3-34   Beveled corner box.

Side-Bracket Boxes

The side-bracket box (Figure 3-35) is a square or rectangular type and is so called since it has a bracket on one side for mounting the box. The side bracket can be to the left or right of the box. Side brackets are generally used only in new installations, since it is difficult to install the brackets once the walls are completed. Side brackets are convenient for mounting boxes on wooden crossbeams in an attic or basement.

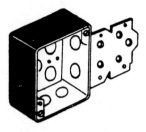 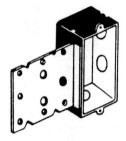

Figure 3-35 Side-bracket boxes. The boxes can be square or rectangular.

Since these areas are exposed, the side brackets can be used in either new or old homes in such locations.

Through Boxes

These boxes are used for installing switches or receptacles on both sides of a wall. Unlike ordinary boxes this type is open both front and back so can be used for switches and receptacles. The advantage of using a through box is that it is not only convenient but economical.

Weatherproof Boxes

Use these boxes for outdoor installation. The distinguishing characteristic of these boxes is that they are equipped with weatherproof snap covers. The covers are spring loaded ones that must be lifted to permit access to the receptacle housed in the box. When the connecting appliance plug is removed, the cover can be snapped shut, protecting the receptacle against the weather.

Round Boxes

This type may sometimes be seen in electrical supply stores but isn't ordinarily used in home wiring. Because of the curvature of the box, it is difficult to fasten input and output power cables to the box, hence, special cable clamps are required. Round boxes are sometimes used for the installation of ceiling lamps.

Box Language

A junction box is one that accommodates wires which are joined to each other. The cover plate of a junction box is intended to enclose these wires completely and so has no openings for receptacles or switches. Junction boxes are available in various sizes, but their function remains the same. Boxes designed for switches and receptacles are sometimes incorrectly referred to as junction boxes.

Ceiling Boxes

These are available in a number of different forms and are dependent on ceiling construction. If the ceiling is made of cement, a masonry drill will be needed for producing holes for lead anchors. The box can then be screwed to the ceiling using lead anchor screws. If the ceiling is wood, the same procedure can be followed by using a wood drill. For plaster or dry wall ceilings, plastic anchors can be used.

Some ceiling boxes have flanges that permit the box to be fastened to the wood ceiling with nails, which are then hammered into position. Special care is needed with plastic boxes, since they can be damaged by an accidental hammer blow.

Another ceiling type uses an adjustable bar hanger. The hanger can be screwed or nailed to a joist. The advantage of a hanger is that the box it supports can be moved to any distance between the joists. The adjustable bar hanger is suitable for attic or basement installation, since the ceilings are open construction. They are not practical when ceilings have already been installed.

Shallow Boxes

In some instances the space behind a wall may not be enough to accommodate an electric box. In that case a special shallow box can be used, but this type can only permit a single switch or receptacle with just one power cable.

Number of Power Cables

The number of power cables that can enter and exit an electrical box is determined by two factors: the volume of the box and the gauge of the wires used. Thick wires not only take up more room, but joining them requires larger wire nuts.

The chart in Table 3-1 shows the number of wires that are permitted in an electric box. The smaller the box, the less its cost. It may also be easier to install a smaller unit. A larger box, however, permits expansion of the wiring system in anticipation of increased future electrical needs.

Dimensions supplied in Table 3-1 are in inches. Ground wires are not included in the count. Boxes are listed as ceiling, utility and in-wall switch and receptacle. Utility boxes are for electrical devices such as washing machines, dryers and ovens.

TABLE 3-1 ELECTRIC BOX SELECTION GUIDE

Type	Box sizes (in.)	Maximum number of wires (excluding ground wires)			
		No. 8	No. 10	No. 12	No. 14
Ceiling	4 × 1½ round/octagonal	5	6	6	7
	4 × 1½ square	7	8	9	10
	4 × 2⅛ square	10	12	13	15
	4¹¹⁄₁₆ × 1½ square	9	11	13	14
	4¹¹⁄₁₆ × 2⅛ square	14	16	18	21
Basic switch or receptacle	3 × 2 × 1½	2	3	3	3
	3 × 2 × 2	3	4	4	5
	3 × 2 × 2¾	4	5	6	7
	3 × 2 × 3½	6	7	8	9
Utility	4 × 2⅛ × 1½	3	4	4	5
	4 × 2⅛ × 1⅞	4	5	5	6
	4 × 2⅛ × 2⅛	4	5	6	7

THE SERVICE BOX

The service box containing either fuses or circuit breakers is the beginning of all electric power in the home. In an apartment the service box may be an in-wall mounted unit, possibly in the kitchen or in some closet. In a house it is usually located in the basement.

Electric power comes into the home via heavy cables, which pass through a meter that measures electric power consumption. The purpose of the service panel is two fold: the separation of the input electric power into a number of branches, and the protection of power in those branches by fuses or breakers.

The Need for Branches

The purpose of electric branches is to divide the tremendous amount of electrical power input into smaller segments that can be used more conveniently. In turn, these smaller segments permit the use of wires that can be manipulated with greater ease.

Another convenience of the branch concept is that each can be handled as a separate entity. As a result, a fault in one branch means that power through all the other branches isn't disturbed.

The use of branches makes electrical servicing easier. Once a problem has been traced to a specific branch, that branch can be shut down by removing its controlling fuse or by turning off its circuit breaker. Narrowing the location of a fault to a branch makes it much easier to solve the problem. Working on all the branches at the same time would be a "needle in the haystack" approach.

The use of branches makes it much easier to protect these power lines by fusing or breakers.

Not all branches are identical in terms of their ability to supply electric power. Power-hungry devices, such as electric stoves, washing machines and their dryers, can have a higher voltage input, supplied by a branch dedicated to each use.

IDENTIFYING BRANCH CIRCUITS

The owner of a home or the renter of an apartment is seldom supplied with an electrical wiring diagram, although such information can be extremely helpful. Fortunately, there is an easy technique for determining the wiring plan.

Electrical power from a utility is brought into the home through an arrangement of three wires reaching from a utility pole. The primary voltage from the utility is 2400 volts and is supplied to a step-down transformer that is pole mounted. The stepped-down voltage is supplied by three wires, with two of these carrying 120 volts and a neutral or zero voltage wire.

Voltage is always measured between two conductors, one of which is the reference or zero, with the other at some voltage, such as 120. The three wires from the secondary of the pole transformer consist of two hot wires and a neutral. The neutral is the reference or common wire for the two 120-volt wires. Thus, 120 volts can be measured between either of the two 120-volt wires and the neutral. The amount of voltage between the two 120-volt conductors is 240 volts. Thus, these three wires can supply a pair of 120-volt lines or a single 240-volt line.

These wires then enter a kilowatt-hour meter, which supplies data on the number of kilowatt hours used. This meter can be mounted on the side of the home or may be inside the home, usually in the basement. From the meter the wires lead into an electrical service box, also known as a fuse or circuit breaker box. Two of the three entry wires are connected to a pair of main fuses followed by a main or master disconnect switch.

The Master Disconnect Switch

In older homes the master disconnect switch, located outside the service box, consists of a handle that can be operated vertically. Subsequently, the master switch has been incorporated in the service box.

It is essential to become familiar with this switch in the event of an emergency that requires the complete disconnection of all electrical power into the home. Opening this switch removes power from all outlets, lamps and appliances throughout.

The Service Box

The service box contains a number of fuses in older homes or circuit breakers in newer ones. The amount of possible current input is rated in amperes and is usually 100 amperes, although it is 200 amperes if the house is wired for electrical heating.

Branch Voltage

The wires extending from the fuse or circuit breaker box are referred to as a branch, and a number of branches are used throughout the home. A particular branch may supply current to several or more electrical receptacles. These are all wired in parallel (also called a shunt connection). The greater the number of appliances that are plugged in (and are operating), the greater the current demand.

Each branch circuit has its own fuse or circuit breaker. By examining the fuse or circuit breaker, you can determine the maximum current that can be handled by that particular branch. If the load—the amount of current demand required for the operating electrical devices plugged into a branch—is in excess of the current tolerance of the fuse or circuit breaker, these will open and interrupt the flow of current. This can happen with an overload condition or a short circuit in the wiring or appliance.

Drawing Branch Circuits

In the power box you will find a sheet of rather strong paper or cardboard pasted to the inside of the front panel. The ruled lines of this paper are numbered. This sheet should contain valuable information about all the branch circuits in the home, but unfortunately it is frequently left blank by the contracting electrician.

Each fuse or circuit breaker has a number printed alongside, with these numbers corresponding to the various branches. A representative box might contain as many as 32 breakers, with the number of breakers corresponding to the size of the home. A three-room apartment would have far fewer breakers or fuses than an eight-room home. In some instances some of the breakers are not used but are present in the event additional branches are to be wired in.

The fuse box data sheet may also have information about the total amount of available current (such as 200 amperes), the amount of voltage (possibly 120/240 volts AC), the number of phases (1 or 3), and the number of available breakers or fuses (either of which is sometimes called a pole).

The numbers listed on the panel data sheet should correspond to the various branches. Thus, a number such as 10 might be followed by information such as:

10   Living Room; Dining Room

This means there is a branch extending from the service box to the receptacles and lights in these two rooms and that this branch is protected by fuse or circuit breaker number 10. The living room might be the first room to which this branch is connected so is the beginning of the branch run. The dining room gets its outlet and lighting power from a continuation of the wiring to the living room and is the end of the run. Electrical problems in the living room could cause power problems in the dining room and vice versa. That is why it is necessary to know which rooms get power from a particular branch. Thus, in this example, a problem in an outlet in the dining room could result in a loss of lights in the living room.

How to Determine a Run

Start by turning on at least one light in all the rooms, either a ceiling light or a lamp. At the fuse box turn off any one breaker marked 15 A (amperes). Locate the room in which the light or lights are turned off. Make a record of the fuse or breaker number and the room or rooms affected. This test should not only include all rooms, but the basement, garage and attic, if these areas are available. Go back to the box and restore the fuse or breaker. Now repeat this action with all the other 15-ampere fuses or breakers, one at a time.

With that completed, follow the same routine with all 20-ampere fuses or breakers. When this job is finished, you should have a wiring record of all electrical branches going to all rooms.

The power branches go not only to all the rooms, but there are special branches for large current appliances, such as electric stoves, oil burners, and clothing dryers. These use 30-ampere fuses or breakers. Open the 30-ampere fuses or breakers, and check to see which appliance has stopped functioning.

With the completion of this check you should have a complete record of which fuses or breakers control various rooms and appliances. Make sure this information is entered on the data sheet on the service box.

In apartments the fuse or breaker box may be in a closet or in the kitchen. In a home the box may be in the basement, if the house is so equipped, or may be in the garage. With the data sheet placed on the door of the box, you will now be able to identify at once the breaker or fuse for any room or appliance and will have a shortcut for electrical servicing.

Branch Circuit Details

Not all branch circuits are the same in all homes or apartments, but in general they follow the approach discussed next.

Lights, electric clocks, small TV sets, electric pencil sharpeners, and similar equipment are operated from a 120-volt line. The wire size of the branch can be No. 14 gauge; the fuse or breaker is rated at 15 amperes. The maximum total power consumption can be calculated by multiplying the line voltage by the maximum current. 120 × 15 = 1800 watts. Small appliances such as clocks, radios, and so on may carry a label indicating the amount of power consumption. Assuming all the appliances are operating at the same time, it is easy to add the total power usage and in that way to determine how close this branch is to the limit. Be sure to include all electric lights and any devices which may be plugged in only temporarily. This will supply a clue as to why fuses and breakers for a particular branch seem to keep opening regularly.

The 20-ampere branch will use No. 12 gauge wire, and its power-handling capacity will be 20 × 120 = 2400 watts. This does not mean increasing the thickness of the wire from No. 14 indicates that the wire has somehow generated more power. All electrical power is obtained from the service box and is supplied by the electric power utility. The use of a smaller gauge wire—that is, a thicker wire—only means a larger power-handling capability.

A branch circuit using No. 12 wire is intended for more current-hungry appliances, devices such as electric irons, portable electric heaters, electric toasters. Note that these components are those that convert electric power into heat power deliberately, since heat is what is wanted. Other electrical components such as electric light bulbs or electric fans may produce heat, but in that case the heat is just a waste byproduct.

Since electric power is the product of voltage and current, a greater current capability can be obtained by increasing current or voltage. If the voltage is doubled, raised from 120 to 240 volts, the current required can be cut in half. Thus, a 120-volt 15-ampere branch can supply 1800 watts. A 240-volt 7-½-ampere circuit can also supply 1800 watts. The advantage of using only 7-½ amperes is that a thinner wire, easier to handle and less costly, can be used.

An electric dryer, for example, may require 7200 watts. This can be supplied by a 240-volt 30-ampere branch (240 × 30 = 7200). A 120-volt branch would need to deliver 60 amperes to supply the same electrical power, and the thickness of the wire could make such a branch impractical.

FUSES

The purpose of a fuse is to prevent current flow from exceeding a predetermined amount. Fuses not only protect appliances but wiring as well. Properly installed and used, they also reduce the possibility of an electrical fire.

Fuse Ratings

A "blown" fuse is one whose current rating has been exceeded. Fuses are rated by the safe amount of current they can pass, ranging (for electrical work in the home) from about 5 to 30 amperes or more. A common type is the 15-ampere fuse used for protecting No. 14 gauge wire.

Basically a fuse contains a short length of metal designed to melt when a certain amount of current flows through it. Since the melting process can cause the metal to splatter, it is enclosed in some sort of housing, commonly a combination of metal and glass. Fuses are available not only in different current capabilities but in varying kinds of construction as well.

Plug Fuses

The plug fuse is probably one of the most common types. It is made with a screw-type socket, as shown in Figure 3-36, and has a window at the top, covered with a transparent material such as glass, through which the fuse element can be seen. These fuses are available in two different shapes, hexagonal and round. If the fuse is rated at 15 amperes or less, the window has a hexagonal shape; if more than 15 amperes, the round shape is used. However, this isn't standard.

Figure 3-36   Plug fuse.

Plug fuses are available in current ratings of 3, 6, 10, 12, 15, 20, 25, and 30 amperes. When a plug fuse blows it often forms a discoloration—a black smudge against the inside of the window. One way of checking a fuse, other than inspection, is to replace it with a known good fuse. A good technique is to keep several known good fuses as "test" units to determine whether an existing fuse is good or not. Another method is to use a volt-ohm-milliammeter (VOM) with its function control set to read resistance. Connect one test lead to the bottom of the fuse and the other test lead to the metal side of the fuse. The meter should read practically zero resistance.

When Will a Plug Fuse Blow?

Whether or not a plug fuse will blow depends on the amount of excess current, the length of time the excess current condition continues, and the speed with which heat can escape from the fuse. For a short-circuit condition, a fuse will open practically immediately. With an overload condition the fuse may or may not open, depending on how long the overload lasts. As a rule of thumb, if the overload is 50 percent, the plug fuse may open in 1 to 15 minutes. As an example, for a 10-ampere fuse, a current of 15 amperes will cause it to open, not at once, but within the 15-minute period.

Fuses aren't precision devices, and most of them have a 10 percent overload tolerance. A 10-ampere fuse, for example, can have a 10 percent tolerance or 1 ampere. This fuse will be able to handle a current of 10 + 1 = 11 amperes, without opening when the current rises to this amount.

The fact that a fuse blows isn't always an indication of appliance trouble. The fuse may be connected into a branch power line that is being overloaded. As more and more appliances are connected and operated, the current drain increases.

Cartridge Fuses

Cartridge fuses are available in two forms: renewable and nonrenewable. The nonrenewable type consists of a cylinder made of some hard, fiberlike material containing the fuse element. This is a short length of metal that melts at a predetermined value of current. It is either soldered to or mechanically fastened to a pair of metal ferrules, one at each end of the fuse housing. The ferrules are actually end caps and are used for clipping the cartridge fuse into a pair of spring metal holders.

Cartridge fuses are generally designed to handle larger currents than plug fuses, but you can get them in current ratings of 3, 6, 10, 25, 30, 35, 40, 50 and 60 amperes. Up to 30 amperes the fuses are 2 inches long, and for currents between 30 and 60 amperes, they measure 3 inches.

Some cartridge fuses are renewable. The fusible metal strip can be replaced after the fuse has opened. To replace the fuse strip, remove the ferrules, mount the new strip, and then replace the ferrules.

Knife-blade cartridge fuses (Figure 3-37) have much higher current ratings than the ferrule type. These can have current ratings of 65 to 600 amperes. Knife-blade cartridge fuses from 60 to 100 amperes are 7-$\frac{1}{8}$ inches long, while those having higher current ratings are longer.

Figure 3-37   Knife blade fuse (upper); snap-in (lower).

Time-Lag Fuses

In some electric circuits a momentary overload current is a normal condition. A motor, for example, is practically a short circuit across a power line when the motor armature starts to revolve. As it starts moving, it develops a counter or opposing voltage. This voltage opposes the line voltage and in so doing reduces the current taken by the motor to some safe value.

A motor might require 30 or more amperes starting current but in operation will take only 5 amperes. If an ordinary fuse is used, it can open under these working conditions. For circuits using motors, the fuse that is used will probably be a time-lag type. This fuse looks just like any other but is capable of tolerating an overload condition for a longer time.

Type-S Fuses

This is a nontamperable time-delay fuse. It has a separate base that can be positioned in a socket in the fuse panel. This base accepts just one fuse size. Thus, a 20-ampere fuse cannot be made to fit if the base is designed for a 15-ampere type. Unlike ordinary plug-in type fuses, there is no possibility of making an error, deliberate or otherwise, of using a higher-rated fuse in place of one that is lower rated.

Fuse Pullers

Plug fuses can safely be removed by gripping and turning the insulation portion surrounding the window. Cartridge fuses are another matter, since they are strongly gripped by their end clamps. It is easy for fingers to slip and touch the end ferrules, and since these are metal, there is a good possibility of getting a shock. This can be avoided in two ways. Turn off the power at the main switch cutting off all available light, and necessitating resetting all electric clocks. To avoid this, use a cartridge fuse puller of the types illustrated in Figure 3-38.

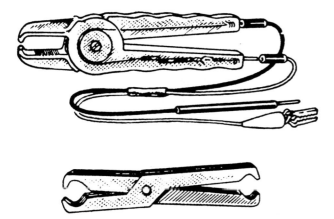

Figure 3-38 Two types of cartridge fuse pullers. The one at the top also has a neon glow lamp and a pair of test leads.

USING THE FUSE OR CIRCUIT BREAKER BOX FOR SERVICING

The main function of the fuse box is to protect branch power lines in the home, to prevent the start of electrical fires, and to protect electrical equipment. But it can also be used as an aid in locating electrical faults.

If an appliance stops working, follow these test steps to find the fault.

1. Make sure the plug of the appliance is in its outlet. Sometimes the plug is accidentally removed. The prongs of the plug must make firm contact when inserted into a receptacle. If the receptacle is a switched type make sure the switch is in its on position.
2. Check whether there is power at the outlet. Do this by plugging in an ordinary lamp, such as a portable light or a neon tester.
3. If these tests show there is no power at the outlet, go to the fuse or circuit breaker box. The inside panel of the box should indicate which fuse or breaker controls the circuit to the inoperative appliance. If the fuse is a plug type, examine it for evidence of a current overload.
4. If the plug fuse has a window that is badly smudged, the overload is severe. This could also be an indication of a short circuit, representing a maximum overload condition.
5. If a fuse has a window that is clear but whose fuse strip is melted, this is evidence of an overload condition, but the overload is a light one.
6. Whether the overload is light or heavy, the next step is to find the cause. Remove all appliances connected to the fused line. Replace the fuse with one known to be in working condition. If the fuse opens, this is an indication of a short in the branch line. The fault may be due to one of the receptacles in that branch. Open each receptacle in turn, and examine the wiring and the interior of the electric box for any smudging. There may even be a telltale odor of burning. If an older, fabric type electrical tape was used, it may have unraveled. Also, an electrical switch may not be correctly positioned in its metal box.
7. If the fault is not in the branch power line and the new fuse did not open plug in each appliance, in turn. The appliance that opened the fuse has some defect.

If none of the appliances caused the replaced fuse to open and all the appliances are functional, there are two possibilities: (a) One of the appliances caused a momentary overload. This is possible if the appliance is motor operated. (b) The load on the line is at a maximum or a little above it. Accumulation of heat inside the fuse could then cause it to blow. The cure is to reduce the load on the line by connecting one of the appliances to some other line—that is, redistributing the load—or by rewiring the branch line with a heavier gauge of wire. Cube taps and strip outlets can result in open fuses if these receptacle devices are used without regard to the current capability of the branch power line.

CIRCUIT BREAKERS

A circuit breaker is a combined fuse and switch. Unlike plug- and cartridge-type fuses, no replacement is needed. When a circuit breaker opens (equivalent to a blown fuse), all that is needed is to correct the fault and to reset the breaker. This is done by depressing the proper breaker switch.

Circuit breakers are also available with a time-lag feature so they can also be used with motors having a high starting current and a lower operating current.

Like fuses, circuit breakers are constructed in various ways. They may be thermal (heat operated), magnetic, or work on a combination of thermal and magnetic properties.

The thermal breaker has a bimetallic element made by bonding two different metals to each other. Each of the metals has a different temperature coefficient of expansion. This means the two metals are affected differently by increases in temperature, and one will expand more rapidly than the other. But since the two metals are joined, a rise in temperature will make the metals bend. The bimetallic element will then act as a latch, tripping the circuit breaker and causing it to open. The bimetallic element will do this when the current exceeds a predetermined value.

Once the breaker is open, current stops flowing, since the breaker, working as a switch, is now also open. Since current no longer flows through the bimetallic element, the element resumes its normal position. The switch, whose external structure is somewhat like an ordinary light switch, can now be reset to its closed position. Current will once again pass through the bimetallic element. If the condition that caused the excessive current flow has been removed, the circuit breaker will remain closed. If the excessive current condition still remains, the circuit breaker will open once again.

The thermal-type circuit breaker is the one most often used in the home, since it is easily capable of handling typical household current.

A magnetic circuit breaker takes advantage of the fact that when a current of electricity flows through a coil, that coil becomes an electromagnet and becomes capable of attracting a bit of ferrous metal. That metal could be part of a switch. As long as current through the magnetic circuit breaker coil remains normal, the magnet isn't strong enough to pull the metal strip away from its closed-switch position. In the case of an overload or short circuit, the current through the magnetic circuit-breaker coil increases substantially. The coil becomes a stronger magnet and attracts the metal strip to itself. This metal strip is part of a switch, and in moving toward the magnet, opens a circuit.

Magnetic circuit breakers are used in applications in which fairly heavy currents flow and are designed specifically for stronger circuit conditions than the thermal type. The thermal type is more sensitive to and more responsive to smaller currents.

The Thermal-Magnetic Breaker

This combination of thermal and magnetic principles is used for installations in which there is a wide range of current flow. The thermal element protects against current surges and overloads in the lower range, while the magnetic element protects against higher current shorts.

Advantages of Circuit Breakers

Breakers have a number of advantages over fuses. There is no possibility of tampering with the thermal or magnetic element, since they are sealed. It is easy to reset a circuit breaker by depressing a switch. Once the circuit breaker is installed and functioning, there is never a need to buy and replace fuses. There is also no possible danger of shock, since the voltage connections are shielded. New home installations use a circuit breaker panel rather than plug- or cartridge-fuse arrangements. When older homes are rewired, it is advisable to replace an existing plug- or cartridge-fuse type with a circuit breaker.

Most home-type circuit breakers have a main breaker used for disconnecting power to the entire home. The main breaker is then followed by a series of breakers, one for each line. Each branch circuit must have a breaker to protect each ungrounded conductor. Therefore, a circuit breaker box can have 10, 20, or more circuit-breaker switches.

What if the Circuit Breaker Opens?

Follow the same procedure as that used in connection with a plug fuse. As a first step, reset the breaker. It may have tripped because of a momentary overload. If this doesn't help, remove all appliances from the receptacle that is breaker protected. Reset the circuit breaker and insert a test light that is in good working order into a breaker-protected receptacle. If the lamp lights, the entire line from the circuit breaker to the receptacle is in good working order. If the lamp does not work and if the breaker persists in opening, turn off its controlling switch, then open and examine each of the receptacles in that line, in order.

Plug-Type Circuit Breakers

Older homes use plug-type fuses, but it is possible to have the convenience of a circuit breaker by replacing the entire fuse panel. This can be expensive and requires the services of a professional electrician.

An easier way is to replace plug fuses with mini-circuit-breaker types. These look like plug fuses and fit directly in place of a plug fuse. They have a reset feature consisting of a small pushbutton in the center of the fuse. When the breaker opens, it can be reset by depressing the pushbutton.

CIRCUIT BREAKERS VS FUSES

Circuit breakers are not only more convenient to use than fuses but have a number of other advantages.

With certain types of fuses it is easy to replace one with another having a higher current rating or even one having a lower rating. The higher rating endangers the branch from the fuse to the load (lights, appliances, and so on). It is also the equivalent of losing the protection supplied by the fuse. The substitution of a lower-rated fuse means the fuse will open with smaller loads, requiring more frequent fuse replacement. Always replace a fuse with one having the same current rating.

Circuit breakers are safer, since their reset handles are made of plastic. While plug fuses are glass enclosed, the glass is very close to the metal body of the fuse. Cartridge fuses are even more risky, since their round ends are metal.

Replacing a Fuse

- Do not replace a fuse without wearing dry shoes or sneakers.
- Use one hand only, with the other hand in your pocket. Your hand should be dry.
- Do not stand on a wet floor. Use a mat, scrap piece of rug, or any other insulating material.
- Use a fuse puller for cartridge-type fuses.
- Make sure you have adequate light. Use a flashlight or a candle if no flashlight is available. Never work on a fuse box in the dark.
- Keep a stock of fuses in some convenient location near the fuse box. Make sure you have enough to replace all fuses—that is, those having different current ratings.
- Do not try to remove the fuse panel. If a fuse has opened, the fault is either in the branch wiring, some receptacle connected to that branch, a lighting fixture, some device plugged into a receptacle, or in the branch wiring. Finding the fault is a process of elimination.
- You might consider replacing the fuse box with one that uses circuit breakers. Use the services of a professional electrician for this.

Evidence of Overload

It is possible to have an overload condition that is close to the maximum current capacity of a branch. If a light is turned on, it may be dim, or a vacuum cleaner plugged in and turned on may not clean properly. The reason for the dim light or poor working condition of the vacuum cleaner is the increased voltage drop in the branch line leading from the fuse box to the power receptacle.

SWITCHES

A tremendous number of electric components—wall plates, receptacles, wires—come in a variety of types. Switches are no exception. Some switches have small handles; others work by the touch of a finger. Switches are designed and manufactured to fulfill many specific functions, to provide various convenience factors as well as meet safety needs and circuit requirements. Even installation needs determine some switch characteristics: flush with the wall, surface mounted, or part of line cords.

Types of Switches

There are two basic types: AC switches, used with alternating current only, and AC/DC switches, designed to work with either alternating or direct current.

Since DC is no longer commonly used, AC switches make up about 95 percent of the applications. AC switches are constructed of lighter materials, fewer moving parts and are smaller in size than a DC switch with equal power capabilities.

The maximum physical size of a general-use switch is governed by the Underwriters' Laboratories, Inc. (UL). How small a switch may be is determined partly by safety restrictions that UL places on the amount of space separating live from dead metal parts. Another greatly varying factor is the electrical capability of a switch, known as the switch rating.

Switch Ratings

A general-use AC switch is rated in terms of amperes and volts. Figure 3-39 shows that each switch has voltage and amperage ratings stamped on its metal bridge or mounting strap. While an AC switch carries just a single ampere rating, it may be rated for either a single voltage or for two voltages. A switch may be marked with a rating of "15 amperes, 120 volts AC." This means it must not be used with voltages higher than 120 nor with equipment that draws more than 15 amperes of alternating current. As another example, a switch could be rated as "15 amperes, 120 volts or 240 volts." This indicates that the switch can be used to control an alternating current of no more than 15 amperes in either a 120-volt or a 240-volt circuit.

Typical Switch Ratings

Table 3-2 lists the current and voltage ratings of various switches.

AC/DC T-Ratings

A T-rated designation applied to certain AC/DC switches means that at a specified voltage and amperage the UL declares a switch safe for controlling circuits in which there may be tungsten lamps (incandescent lamps). This extra safety consideration

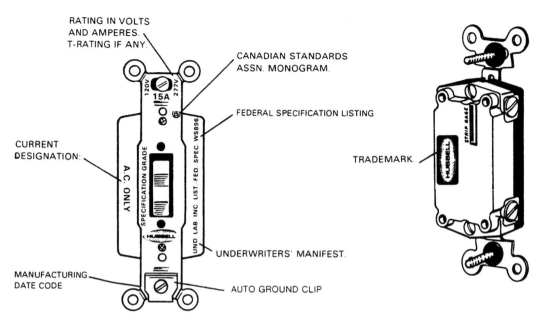

Figure 3-39 Inscriptions of a typical switch. Note rating legend in volts and amperes on the bridge. (*Courtesy Hubbell Incorporated, Wiring Device Division*).

TABLE 3-2 SWITCH AMPERAGE AND VOLTAGE RATINGS

15 amperes, 120 volts
15 amperes, 120 or 277 volts
15 amperes, 208 or 277 volts
20 amperes, 120 volts
20 amperes, 120 or 277 volts
20 amperes, 208 or 277 volts
30 amperes, 120 volts
30 amperes, 120 or 277 volts

is needed because in the first $\frac{1}{240}$ second after a switch has been closed, incandescent lamps have an initial flow of current up to 8 to 10 times the normal load the switch would handle.

Instead of being rated for only one amperage, AC/DC switches are rated for two, depending on what voltage the circuit carries. For example, by looking at the listing for an AC/DC switch you might find:

	Amperes	
Description	125 V.T.	250V
Single Pole	10	5

What this indicates is that the switch can be used to control an incandescent lamp load of 10 amperes in a 125-volt circuit or a current of 5 amperes in a 250-volt circuit.

Since all AC switches are capable of withstanding the initial surge of power of incandescent lamp loads, switches exclusively for alternating current (AC switches) need not carry the T marking.

One important fact to remember when considering AC/DC switches with two voltage ratings is that the switch may be T-rated for the lower voltage and higher amperage but not for the higher voltage and lower amperage. This is emphasized by including the letter T following the lower voltage.

Grade Designations

Another general term you may see or hear applied to a switch is specification grade. This means the manufacturer has classified it as the highest quality switch he produces. Because there is no unanimously accepted definition of this term, it does not indicate that the switch is equal in quality to any other manufacturer's specification grade switch.

Other terms commonly used but not specifically defined are residential grade and intermediate grade.

In addition, switches are also defined by the National Electric Code as *general-use snap switches* for branch circuits. The word snap is somewhat misleading because this classification includes all types of switches—key operated, toggle, and so on. As used here, it indicates that after the switch actuator (button, toggle or whatever) has been moved some amount, the contacts "snap" over to the other position.

Switches for Motors

General-use AC switches cannot be horsepower (hp) rated. They can, however, be used to control motor loads up to 80 percent of the ampere rating of the switch. So an AC switch rated for 20 amperes can be used for a motor load of 16 amperes.

Table 3-3 is a listing of switch current capability in terms of hp.

HOW SWITCHES WORK

A switch is not a load and does not require current or electrical power. A switch is always put in series with a power line, never in shunt with it. A load does require electrical power and is always put in shunt with the power line. Switches and loads can be considered as opposites. Switches are control devices determining the on or off time of a load.

TABLE 3-3 CURRENT/HORSEPOWER CONVERSIONS

	AC General Use Snap Switches	
	120 Volts	240 Volts
15 Amps	½ HP	2 HP
20 Amps	1 HP	2 HP

	AC Motor Control Switches	
	120 Volts	240 Volts
30 Amps	2 HP	3 HP

The Single-Pole, Single-Throw Switch

The single-pole, single-throw (SPST) switch is the simplest, and quite possibly, the most widely used of the entire switch family. The SPST switch in Figure 3-40 is produced commercially, but the type used in the home is much more compact and is completely enclosed. Electrically, the two switches are identical.

    White neutral wires are through wires and are not connected to any of the switch terminals. Only black wires are switch connected. When the switch toggle is in its up position it is closed and current flows through the switch. When in the down position the switch is open and current flow stops. The words on and off may be marked right on the toggle. The on condition is sometimes called the make position; the off condition is called the break.

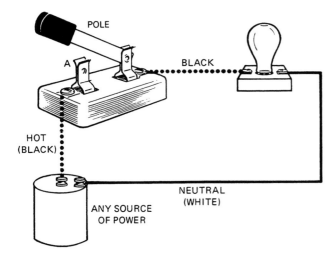

POLE

BLACK

HOT
(BLACK)

NEUTRAL
(WHITE)

ANY SOURCE
OF POWER

Figure 3-40 Action of a single-pole, single-throw switch. (*Courtesy Hubbell Incorporated, Wiring Device Division*)

No damage will be done if the switch connections are transposed. The setting of the toggle will be reversed.

The switch shown here consists of a hinged copper blade used to make or break contact between a pair of metal terminals, with the entire structure mounted on an insulating base. When the switch blade is in its up position, the circuit is interrupted and no current flows to the load, a lamp in this example.

The movable switch blade is referred to as a *pole,* and since there is just one of these, it is called a *single pole* (SP). This pole can move in just one direction and so is called a *single throw* (ST). The switch is designated as a single-pole, single-throw (SPST) type.

The switch is in series with the power line with one terminal connected to a hot lead, and its other terminal also connected to a hot lead. In effect, the hot lead is broken open to permit the insertion of the switch.

The connecting terminal A is used as the make/break contact point of the switch.

When the switch is closed, one side of the load is connected to one side of the power source; the other side of the load is connected to the other side of the power source. Thus the load is in parallel with or in shunt with the power source.

Current Flow

All current that leaves a power source, no matter how many loads or closed switches it may flow through, must return to that source, and for that purpose it requires a pair of conductors. In the case of the circuit of Figure 3-40, the source is shown as a battery. Current leaves the negative terminal, flows through the neutral wire, through the load, through the switch (when it is closed), arriving at the positive terminal of the battery. It flows through the battery from its plus electrode to its minus electrode and then continues its movement through the neutral wire.

If the power source is replaced by an alternating current (AC), the current circulates back and forth, first flowing in one direction, then the other. It is important to remember that the same amount of current flows in all parts of the circuit and that the current flows through the neutral wire as well as the hot lead.

The Double-Pole, Single-Throw Switch

The double-pole, single-throw switch in Figure 3-41 is equipped with a pair of poles. However, its blades move in just one direction; that is, it has a single throw. Because of these physical characteristics it is a double-pole, single-throw (DPST) type.

Figure 3-42 shows there are two loads controlled by a single-pole, double-throw (SPDT) switch. One of these is a lamp, the other an electric bell. Two things happen when the switch blade contacts point D. Current can now flow through the electric bell (load 2); current cannot flow through the lamp (load 1). When the switch blade is moved so it contacts point C, load 2 is disconnected while load 1

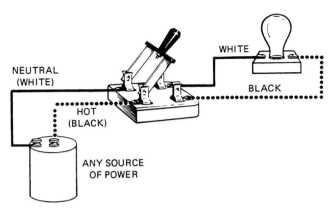

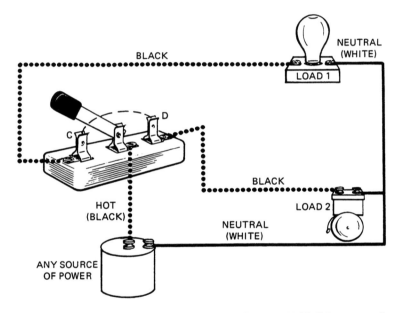

Figure 3-41 Double-pole, single-throw circuit diagram. (*Courtesy Hubbell Incorporated, Wiring Device Division*)

Figure 3-42 Single-pole, double-throw switch. (*Courtesy Hubbell Incorporated, Wiring Device Division*)

receives electrical power. The same result could be achieved by using a pair of SPST switches, but using the DPST makes sure that the two loads do not operate at the same time.

The Double-Pole, Double-Throw Switch

The double-pole, double-throw (DPDT) switch in Figure 3-43 closely resembles the SPDT switch of Figure 3-42. It is actually a pair of SPDT switches mounted side by side and insulated from each other. In this diagram either load 1 or load 2 can

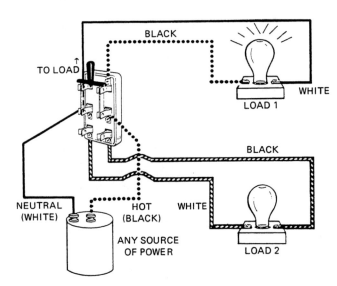

Figure 3-43 Circuit arrangement using a double-pole, double-throw switch. (*Courtesy Hubbell Incorporated, Wiring Device Division*)

be made to function, but both cannot do so at the same time. Electrical power is connected to the center terminals of the switch. When the switch is in its upper position, electrical power is supplied to load 1; when in its down position, power is delivered to load 2.

Three-Way Switches

The purpose of a switch is to permit the delivery of power to one or more loads. It must also be able to interrupt the delivery of that power.

The circuit in Figure 3-44 uses a pair of SPDT switches. With this arrangement it is possible to control a single load from two different locations and permit power to be directed along a choice of routes.

In this illustration when blade A is on contact X and blade B is on contact Z, current flows to the light. The light can then be turned off from either location by moving either blade to the opposite contact.

There are two on and two off settings:

ON: AX–BZ and AW–BY

OFF: AX–BY and AW–BZ

The illustration shows the two switches adjacent to each other, but this is done to simplify the diagram. One switch might be at the lower level of a flight of stairs; the other at the top of the stairs, with the controlled lamp either up or downstairs.

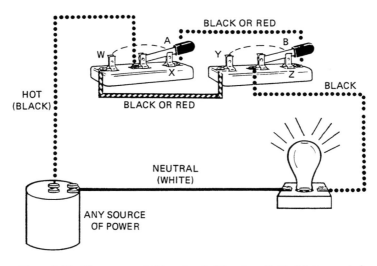

Figure 3-44 Three-way switching circuit. (*Courtesy Hubbell Incorporated, Wiring Device Division*)

Four-Way Switches

A four-way switch, Figure 3-45, is a special type of DPDT switch. What makes it special is that each contact is connected by a jumper to the diagonally opposite contact, as shown in the rear view of the drawing. These jumper wires are insulated from each other.

When the handle is in the up position, current flows through the blade, reaches

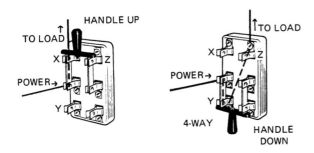

Figure 3-45 Wiring of a four-way switch. (*Courtesy Hubbell Incorporated, Wiring Device Division*)

contact X and continues to the load. When the handle is down, current flows to contact Y, then through the jumper to Z, then out to the load.

Consequently, a four-way switch makes it possible to route current to either of two conductors.

The four-way switch is a multiple control type. Thus, it can be used to control a load, such as a light (Figure 3-46) from three locations. The drawings show two possible wiring arrangements using a pair of three-way switches and one four-way switch.

Electrical power is supplied to point Z at location 1. Current flows from that point to the right center contact of the four-way switch. When the four-way's handle at location 2 is in the up position, it routes the current to one contact of another three-way switch at location 3. Since the latter's blade is on this contact, the current flows through to the light.

Turning the four-way's handle to the down position turns the light out because the current is now directed to the other contact of the three-way switch. Since the three-way blade is not on this contact, current cannot reach the light.

By changing the handle position of any of the three switches, the light can be turned on again.

A load can be controlled from four locations by putting two four-way switches between a pair of three-ways, from five locations by placing three four-ways between a pair of three-ways, and so on.

Four-way switches are available in various types: toggle, locking and press switch.

Toggle Switches

A toggle switch is one that is operated by moving its handle (called a toggle) up or down. Most toggles are *maintained contact* types. This means the handle will remain in its selected position when moved up to its on position or down to its off position.

Toggle switches are available in a variety of forms: illuminated (the handle lights when the load is turned off), pilot light (the handle lights when the load is turned on), and grounded types in which a ground lug is provided on the switch to ground metal wall plates when required.

Lock-Type Switches

This is a switch that is operated by using a removable metal key that fits into a keyway in the face of the switch. Its purpose is to prevent unauthorized persons from turning the controlled electrical device on or off. This type of switch is generally used in schools, hospitals and prisons. It can also be used in the home to keep unwanted electrical use out of the hands of children.

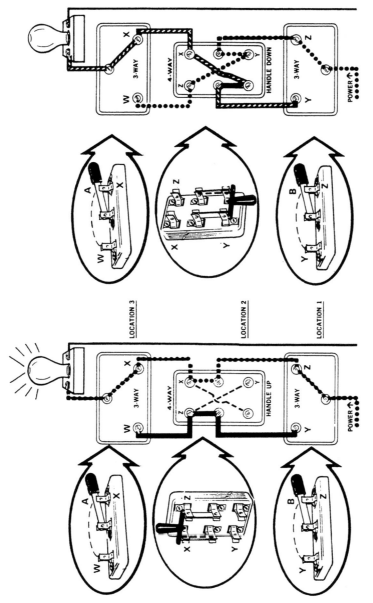

Figure 3-46 Circuit arrangements for three-location control. (*Courtesy Hubbell Incorporated, Wiring Device Division*)

125

Momentary-Contact Switches

A momentary-contact switch remains in its on or off position only as long as its toggle is held there by hand. A doorbell pushbutton is an example of the momentary-contact principle.

Press-Action Switches

This switch can be recognized by the shape of its toggle, which consists of a sloping, wedge-shaped nylon button. A light push of the finger or elbow is sufficient to drive a cam that makes or breaks the switch contacts. These switches are for use with AC circuits only.

Some press-action switches contain a tiny glowlamp that lights the button, but glow lamps aren't restricted to this switch type. In pilot-light types the button glows red when current flows through the load, advantageous since it helps monitor loads that are remotely located. Another switch type has a button that glows only when the load circuit is open.

Press-action switches are available as single- or double-pole types and as three- or four-way units. They are rated for 15 or 20 amperes and for 120 or 120-277 volts AC only.

THE LANGUAGE OF SWITCHES

Switches can be categorized in a number of ways. They can be classified as AC or AC/DC, in terms of voltage and current, or by their poles, throws and ways. They can be designated as single-pole, single-throw; single-pole, double-throw; double-pole, single-throw; double-pole, double-throw; three-way and four-way.

Make and Break

The number of poles is used to designate the number of wires acted on; that is, a single-pole switch will make or break only one wire in a circuit. By contrast, a double-pole switch will make or break two wires in a circuit. Since a switch is intended to permit current to flow or not to flow, *make* indicates that the switch has made current flow continuous; *break* is the opposite and means that the movement of current has been interrupted.

Throws

This is the number of contacts that can be made to a common terminal. Thus, a single-throw switch can control only one set of contacts and one circuit, whereas a double-throw switch can control two circuits.

Ways

This is the number of locations from which a load can be controlled. A three-way switch combination allows load control from either of two locations. Inserting a four-way in between two three-way switches permits load control from either of three locations. For each four-way added between the two three-ways, the control locations increase by one.

Load

A load is any device that receives current. A lamp is a load. So is a toaster or broiler or motor. A load can be *heavy;* that is, it requires a substantial amount of current. Also, it can be *light,* needing only a small current. The terms are relative. A 10-ampere motor, ordinarily considered a heavy load, is a much lighter load than a motor requiring 25 amperes.

There are two types of loads that are extremes. A load that is disconnected from the power line by a switch does not use current and is a zero-load condition. A short circuit, a condition in which the two current-carrying wires (the black and white leads of a branch power line) touch each other, is the heaviest possible load and demands the maximum current from the source.

Three-Level Touch Switches

These are completely unlike the usual switches intended for lamps, since they are electronic rather than mechanical. They are solid-state devices and are available in two versions: rectangular, for simple mounting in limited space, or circular for easy lamp mounting. When installed, all they require is a light touch on any metal lamp surface. A first touch produces 30 percent, a second touch 70 percent and a third touch results in 100-percent lamp brightness. The circuitry built into the touch switch unit compensates for lamp or fixture size. It is not polarity sensitive at the line cord leads, permitting the use of any type of plug, nonpolarized using equal-sized prong plugs, polarized with unequal prong sizes or three prong grounded types. In Figure 3-47 the touch lead to the switch is color coded blue. The unit is

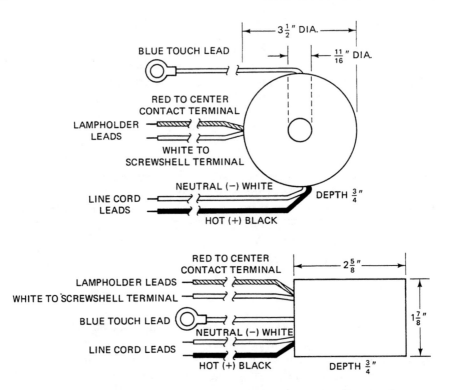

Figure 3-47   Electronic touch switches. (*Courtesy Leviton Manufacturing Co., Inc.*)

rated for 200 watts, 125 volts AC only, is designed for RF (radio-frequency) protection, and should not interfere with radio or television reception.

GROUND FAULT CIRCUIT INTERRUPTER

The current that flows from a service box to an appliance circulates back and forth along the black and white wires, and it does so since the current is AC or alternating. While the black wire is called "hot" and the white "neutral," the same amount of current should flow through both. If, because of some fault in the electric circuit, more current flows through the neutral wire, that current will try to find its way to ground.

The neutral wire is connected to the separate ground lead, and that ground lead is connected to earth ground. So the path for current through a neutral wire is shared by the ground wire and also by a path through the ground. Some excess current in the neutral wire will take advantage of the ground wire and earth connections. Touching a neutral wire, or a ground connection when this excess current condition exists, means an additional path has been provided. And this is where danger exists.

Ground faults happen when electric current leaks to the ground from electrical devices. This can happen anywhere, but especially in areas that are subject to moisture. Inside, that means bathrooms, the kitchen, basement, workshop, and laundry room. Outside it means around the pool (especially for plug-in motors), bug zappers, electric barbecue grilles and charcoal lighters, all exterior plug-in lighting, Christmas lights, and all power tools, including electric hedge clippers, saws, drills, paint sprayers and all electric lawn and garden power tools.

The current that flows through the black and white wires is often in terms of amperes. The excess current in the white wires can be a small fraction, possibly one or two tenths of an ampere. The human body, though, can tolerate even less than this, and only for a very short time, possibly just one or two seconds. The result can be electrocution.

A device called a ground fault circuit interrupter (GFCI) or sometimes just a ground fault interrupter (GFI) can be used to stop the flow of this tiny leakage current and it can do so practically immediately.

SHOULD ALL APPLIANCES BE GROUNDED?

As a general rule, it is advisable to ground appliances, but there are exceptions applicable to electrical devices that use a heating element that can be reached with the fingers, or some metallic tool. These are appliances such as toasters, broilers and electric heaters. In these appliances the heating element is directly in parallel with the power line after they have been turned on. If the outer metallic housing of such appliances is grounded, there would be a direct path from the hot to the cold side of the power line. Anyone touching the heating element and the metal housing could receive a shock.

The precaution is simple. Never—but never—touch the inner heating mechanism with a fork while the unit is plugged in. If a slice of bread gets caught in a toaster, turn the toaster off by removing its plug from its receptacle. Do not—repeat NOT—depend on an on-off switch or an automatic turn-off mechanism. Work on the toaster, or broiler, or electric heater only after its plug has been pulled out.

Removing a section of locked-in toast from a toaster isn't uncommon. Somewhere, somehow, the idea got started that toast isn't an electrical conductor. If the toast has become compacted, it will permit the passage of a leakage current from the heating element to the metal casing. Matted crumbs in a live toaster are an ever-present danger, since they can be electrically conductive.

4

The "How To" of Home Wiring

The preceding chapter described the electrical parts commonly used in home wiring. Basically, these parts are few in number; it is their variation that is so substantial. The parts consist of fuses, circuit breakers, wires, electric boxes, receptacles, switches, ground fault circuit interrupters, incandescent and fluorescent lights.

Just a single one of these, such as switches, for example, is available in dozens of forms, with each making a different circuit arrangement possible. Sometimes, for a given application, more than one kind of circuit can be used, and then, for a practical installation, it becomes necessary to select that circuit which will provide the maximum economy in copper and will require the least amount of labor.

Some installations are simple and require no more than the substitution of one part from another. Others involve new or additional wiring, and in this case you must determine whether to tap into an existing branch or start a new branch, which wire gauge to use, and which new part would be most suitable. Just replacing an electrical part involves a decision of which to use.

HOW TO PLAN WIRING

For some home repairs no planning is needed. To replace a light bulb, the obvious solution is to remove the defective item and substitute a new or known good light. But even here some planning may be needed, unless an identical replacement is to be used. A decision will be needed on wattage rating and whether the bulb is to be clear or frosted. There are also many different bulb styles to consider.

It is best to start with a sketch before adding more wiring to a room. This will help in determining the best location for receptacles and switches.

Planning is even more important if the wiring is to involve more than one room. Figure 4-1 is a preliminary sketch for a wiring installation. It has been kept deliberately simple. For example, it does not include wiring receptacles, either wall or ceiling types, or junction boxes. These can be added to subsequent sketches.

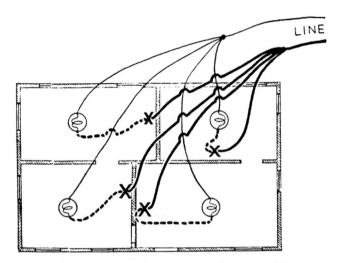

Figure 4-1 Preliminary drawing of a wiring installation.

Start with a pair of lines to indicate the two wires, the black and white, of the power system. Use a thick line for the black wire and a lighter line for the white. Ground wire connections will be needed, but these are usually not included in the sketches.

The drawing shows a first effort. There is no need to draw switches, for you can indicate them by the letter X. Note that the switches have been put in series with the black wires only and that the white wires are all continuous.

Each pair of lines represents a subbranch circuit. Since there are four such pairs, the wiring involves four subbranches. The two wires at the upper right marked "line" represent the branch line, and it continues directly back to the fuse or circuit breaker box, or to a junction box.

The first sketch can be confusing because of the multiplicity of lines and because they cross each other. A better arrangement is the improved sketch in Figure 4-2. Instead of separate neutral wires going directly back to the fuse or breaker box,

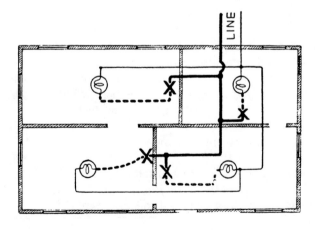

Figure 4-2 Revised drawing of a wiring installation.

the neutral line runs from lamp to lamp, so there is a reduction in the amount of neutral wiring. Also, instead of having separate hot (black) leads, these are now tapped off a single hot wire.

The wiring drawing may include additional information such as wire size, type of wire to be used, fuse or breaker current rating, and total power used, assuming all four subbranches are on at the same time. The symbol for a lamp used in the drawings may represent a light fixture containing one, two or more bulbs. The diagram does not indicate the possible use of dimmer controls. Finally, the diagram does not indicate the path the wiring is to take behind the walls.

The sketch is helpful, even if the do-it-yourselfer decides that this project requires the services of a professional electrician, for it communicates specifically and directly and leaves little chance for misunderstandings. Further, if the branch is to be a new one, it may require connections at the service box, plus the use of a new fuse or breaker, work that is best left in the hands of a professional.

While the preferred technique is to connect the new branch line to the service box, this may not be necessary. If an under-utilized branch line is available, a connection can be made to its junction box. To be able to do this, it is necessary to have a branch line plan; that is, to recognize branch lines, their current capacity, the total number of loads, and the power demands of each of the lines.

The fuse or breaker used by each branch line supplies information that permits calculation of the branch power capability. A fuse rated at 15 amperes connected in a 120-volt line will have a maximum power rating of $15 \times 120 = 1800$ watts. A 20-ampere fuse also connected to a 120-volt line would have a capability of $20 \times 120 = 2400$ watts. Because of its greater capacity the 20-ampere line would be preferable. For a 15-ampere line it would be desirable to limit total used power, with all receptacles active, to 1500 watts; limit the 20-ampere line to 2000 watts. Power demands, however, can be variable, depending on the type of load. Appliances that use motors will have a larger starting current than operating current, so their power demands are higher at their start and decrease as the motors reach operating speed.

THE NEED FOR RECEPTACLES

Every home needs more receptacles, a requirement that can be met in seven different ways.

Cube Tap

This is one of the least expensive, quickest and easiest ways to provide additional receptacles. It is also the least desirable, for not only does it spoil the appearance of a room, but it advertises the fact that the house is inadequately wired. This not only reduces the value of a home but makes it more difficult to sell.

Line Cord Ending in a Cube Tap

This is a temporary device, but sometimes temporary ends up in long-term use. Its advantage is that it permits the extension of a receptacle to some distance. It can be a hazard, since a line cord is easy to trip over. If actually used on a temporary basis, setting it up and then removing it for storage can be a time-consuming nuisance. It is convenient, however, when necessary, but certainly adds nothing to the appearance of a home.

Outlet Strips

These are part of the tap family. Unlike cube taps and line cord taps, they are intended for permanent installation. Some of them are equipped with individual circuit breakers, lights to indicate that all the receptacles are active, and an on-off switch. They are usually mounted behind a desk or in some easily accessible area. They are equipped with three-prong polarized receptacles.

Pull Chain Light Sockets

Some pull-chain porcelain ceiling light sockets used in attics and basements are receptacle equipped. They are satisfactory for these areas of the home, provided they are not overloaded.

On-the-Wall Wiring

This is a fairly easy way of installing new receptacles. The channel containing the wiring can be painted to match existing wall color or can be papered to match the existing covering. While they are not too desirable for living and dining rooms, they are suitable for dens, playrooms, guest bedrooms and basements having wood covered walls.

In-the-Wall Wiring

This is the most difficult and the most expensive method. It usually requires the services of a professional electrician. It may call for repapering, plastering, or repainting part of a room, and possibly more than one room.

High-Fidelity Sound System Receptacles

The individual components in a hi-fi system are AC powered and as such require a connection to a receptacle. To avoid the need for individual power cords leading to individual receptacles, many of the components are equipped with active outlets mounted on rear aprons. This permits one component to get its electrical power from another, so the entire system may be connected to a receptacle with just one cord. However, the receptacles supplied are restricted to use by the sound system.

How to Add a First Floor Receptacle

Adding a receptacle to a room on a first floor, if that floor is directly above a basement that doesn't have a finished ceiling, can be easier than wiring upper floors. Locate a space between studs where the new receptacle is to be installed, as in Figure 4-3. The new receptacle will require its own electric box, and it will need to be supported against a stud. After you have determined the exact location of the new receptacle, carefully measure the distance between it and the positioning of the old receptacle which will act as a junction box.

After you have located the drill areas, use a wood drill and bit to drill the holes. After drilling, shine a light up from the basement through the holes to make sure they are properly located. The holes must be large enough to accommodate the Romex, if that is the cable to be used. The wire gauge of the new cable should be the same as that used by the wiring reaching the original receptacle. Push the cable through the holes to make sure it has the correct length. Before connecting the wires to the original receptacle, turn off the power to it at the service box.

The cable must not be suspended by its own weight. Instead, use cable straps about every three or four feet, fastening the cable to the wooden ceiling of the basement or to any other available wooden support.

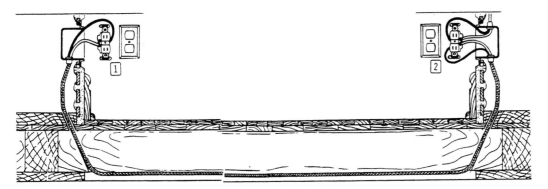

Figure 4-3 How to add a first floor receptacle.

Before mounting the new electric box, determine which knockout to use. Do not use more knockouts than necessary. Mount the electric box against the stud and then use a cable holder as shown in Figure 4-4. Connect the cable to the correct terminals of the receptacle, then mount the receptacle in its box.

Remove the original receptacle from its box, and add the wiring of the new cable. After making these connections, restore the receptacle to its box. Turn on the power at the service box and then check both receptacles, the original and the one that was added.

Electric work to be done in the home can often be done by a single person. When the required wiring is in wall, it is better to have the assistance of a second person, especially someone having some building or wiring experience.

Figure 4-4 Method of fastening Romex to an electric box.

HOW TO INSTALL ELECTRICAL OUTLET STRIPS

There are various ways of increasing the number of power receptacles (outlets) in a home. The easiest way, but not the most desirable, is to use an outlet tap, also known as a cube tap. Since it extends from the receptacle and since it has line cords coming into it from all directions, it is very unattractive and calls attention to the inadequacy of the wiring. It also can contribute to line overloading.

Another method is to install additional outlets, either in-wall or surface types. These involve rewiring, and for in-wall types it means breaking through the wall, with subsequent repainting or repapering needed. It is a good method, but it does take time and a fair amount of work.

Still another technique is to use an electrical outlet strip, a method that could be considered related to the cube tap. The electrical outlet strip, Figure 4-5, consists of a number of outlets, ranging from 4 to 10, mounted in a rectangular-type metal container. All of the outlets are prewired in parallel. Units are equipped with a 6-foot or 15-foot No. 14/3 (three AWG No. 14 wires) heavy duty cord whose free end is equipped with a three prong plug.

The strip may be equipped with a lighted master on-off switch and a circuit breaker with a press-to-reset button. An advantage of the strip is that it quickly supplies receptacles to a room that has an inadequate number.

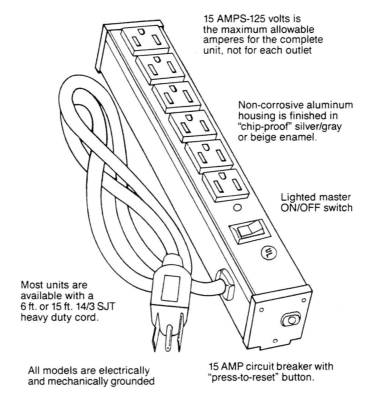

15 AMPS-125 volts is
the maximum allowable
amperes for the complete
unit, not for each outlet

Non-corrosive aluminum
housing is finished in
"chip-proof" silver/gray
or beige enamel.

Lighted master
ON/OFF switch

Most units are
available with a
6 ft. or 15 ft. 14/3 SJT
heavy duty cord.

All models are electrically
and mechanically grounded

15 AMP circuit breaker with
"press-to-reset" button.

Figure 4-5 Essential features of an electrical outlet strip. (*Courtesy Brooks Electronics, Inc.*)

While many outlet strips have master on-off switches so that all the receptacles are either on or off at the same time, there are some whose receptacles can be individually switched and come equipped with separate combination switches and pilot lights so it is easy to determine just which receptacle is live. Some strips are also equipped with surge suppressors.

Outlet strips are usually mounted along a baseboard with the heavy duty line cord plugged into the nearest convenient outlet.

Precautions in Using an Outlet Strip

While the outlet strip supplies additional receptacles, it does not, and cannot, increase the amount of available electrical power in the branch line to which it is connected. If a home is equipped with a 15-ampere and a 20-ampere branch power line, it is better to use the 20-ampere line. Electrical devices plugged into the outlet strip, plus all those connected elsewhere along the branch power line, must not exceed the power capability of that line.

Remote Control of an Electrical Strip

Components that are connected to an electrical strip usually have individual on-off switches. However, some strips do have provision for remote control, as shown in Figure 4-6. The strip is plugged into an available receptacle. The on-off switch of the strip is connected to a two-wire line, which also ends in a switch. The electrical strip can then remain turned on at all times, while the more convenient on-off switch, possibly positioned under a desk top, can be used for control of the strip.

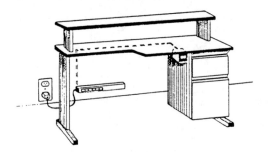

Figure 4-6   Remotely controlled electrical outlet strip. (*Courtesy Brooks Electronics, Inc.*)

HOW TO DO EXTERNAL WALL WIRING

Possibly the most difficult part of wiring (or rewiring) a home is breaking into a wall, fishing wires behind a wall, or trying to run power lines from one room or one floor to another. This may be forbidden by a lease in an apartment, but it can add value to a house.

It is possible to install new switches or receptacles without breaking into walls or ceilings. Do this by running wires from an existing outlet along the surface of the wall to a new switch or outlet. Then cover the wires with a UL-listed plastic channel.

Figure 4-7 shows how to use external wall wiring starting from an existing receptacle. It uses straight channels, flat elbows, outside elbows, an in-channel receptacle, a T fitting, in-channel switch, an inside elbow, and a circular fixture box.

Outside the wall wiring can be used in an easy way to add one or more receptacles to a room. With this method it is not necessary to cut into a wall. Locate the new receptacle, as in Figure 4-8, measure the distance between the old and new receptacles, and cut the wire accordingly. Use Romex and conceal it in the channel that will cover and support the wiring. With power disconnected, connect the cable to the old receptacle and to the new one.

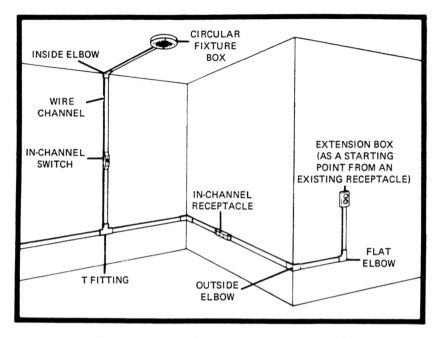

Figure 4-7   External wiring installation. (*Courtesy The Wiremold Company*).

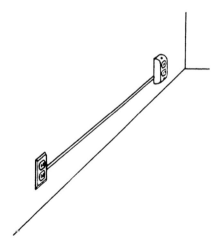

Figure 4-8   Outside-the-wall wiring for installing a new receptacle.

Mapping the Project

Before starting the wiring project, do some planning. Measure the distance from the starting point to the finish. Make sure you have all the parts needed to avoid interrupting the work unnecessarily. Measure the exact distance from the existing outlet to the proposed location of the new switch or outlet. Take note of the planned route—up, down, around corners, onto ceilings, elbows. Finally, make sure you have at hand all the tools you will need. These will include standard and longnose pliers, flat blade and Philips head screwdrivers, wire snips, a pocket knife, a fine tooth hacksaw, a fine tooth file, and a bubble glass level.

How to Proceed

Be sure to turn off the electrical power at the fuse or junction box. As a further assurance plug a test light into the receptacle being used as the starting point. Remove the existing wall plate, loosen the outlet and its receptacle, and install the special base plate intended for this project.

The channel that will contain the wiring consists of two parts: a base and a cover. Run the channel base from the existing outlet to where the new switch or receptacle is to be positioned. To make sure the base is correctly positioned, horizontally and vertically, use a bubble glass. Fasten the base to the wall.

Connect the white, black and ground wires of the new extension cable to the old receptacle. Use No. 12 or No. 14 THHN wire or Romex having the same gauge. Put the full length of wire in the channel, and use channel clips (these are supplied with the channel) to hold the wires in place.

Mount the external receptacle holder. It should fit neatly against the wire channel, making a butt joint. Connect the end of the cable that is in the channel to the new receptacle. Fit the receptacle into its holder. After it has been screwed into position, turn the power on and test the new receptacle for power output. Snap the wire channel cover into place.

The channel can be left as it is or it can be painted with a latex-based paint or stain to match the color of the wall. It can also be papered to match that used on the wall.

HOW TO TEST A RECEPTACLE

There are a number of ways to test a receptacle to determine if voltage is available or if it is properly grounded. You can use a desk or table lamp, assuming it is in working order, with its switch set to its on position. The lamp can be used to check for voltage but not for correct grounding of the receptacle. Another way is to use a test unit known as a volt-ohm-milliammeter (VOM) with its function selector set to AC and its range selector positioned to read at least 200 volts maximum. The easiest way is with the help of a glow lamp, also known as a voltage tester.

Simply insert the test leads of the unit into both slots of the receptacle. The glow of the lamp indicates that the receptacle is live, whether the receptacle is a two-slot nonpolarized type, or whether it has two slots and a ground terminal.

The tester can also be used to check to make sure the receptacle is grounded. For a two-terminal, nonpolarized unit (as in Figure 4-9a) put one test lead on the center screw and insert the other test lead into the two slots, in turn. With one of the slots the lamp will not glow, while with the other it will light. If no light is obtained with either of these tests, the receptacle is not grounded, and it should be removed for further examination.

The same test can be done on a three-slot polarized receptacle, as in Figure 4-9b.

(a)   (b)

Figure 4-9   How to test receptacles for voltage and ground.

HOW TO REPLACE A RECEPTACLE

When working with electrical appliances or devices of any kind, the first step—always—is to turn off the power source. The best way to do this is to remove the fuse or deactivate the circuit breaker controlling the branch circuit being worked on. In addition, always check the receptacle to make sure the power is off. Do this with a test lamp, a test instrument, or a neon bulb tester.

The receptacle, whether in-wall or wall-surface mounted, is covered with a wall plate held in place by a single, centered machine screw. Remove the screw using a flat blade screwdriver. Be careful not to lose the screw, especially if the screw color matches the wall plate.

If the wall has been painted and some of the paint has covered the sides of the wall plate, use a single edge razor blade to cut through the paint. A better method is to use a single-edge razor blade holder. After removing the wall plate, the front surface of the receptacle will be exposed.

The receptacle is held in place by two machine screws, one at top center, the other at bottom center. Remove these screws, using a flat blade screwdriver. After the screws are out, pull the receptacle from its box. You will find a pair of white wires and a pair of black. The black leads (Figure 4-10) are connected to a pair of brass screws on the right side of the receptacle as you face it from the front. The white leads (not shown) are connected to silver-colored screws on the left side of the receptacle.

After the receptacle has been moved out of its electric box, unscrew the black

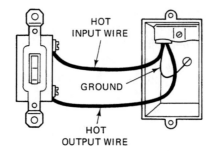

Figure 4-10 Exposed view of receptacle after it is pulled out of its electric box.

and white wires. The screws are usually captive types and can be turned counterclockwise for a limited distance. There may also be a ground wire, either bare or color coded green if insulated. This wire is connected to a ground terminal at the lower left of the receptacle. If it is screw connected, turn that screw counterclockwise to release the wire. In some receptacles the ground wire is clip connected so all that is necessary is to disengage the wire. There may also be a ground wire connected to the receptacle box if it is a metal type. Do not disturb this connection.

If the new receptacle is to be screw connected, the loops at the ends of the white and black wires may need to be opened only slightly. If you prefer using push-in connections (Figure 4-11), straighten each of these wires.

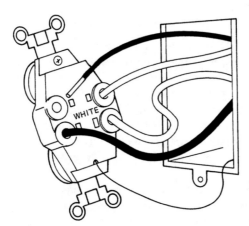

Figure 4-11 Push-ins can be used instead of screw connections.

HOW TO POSITION A RECEPTACLE PROPERLY

Sometimes, when replacing an electrical receptacle box, you may learn for the first time that the box was mounted at some angle inside the wall. The box may be so securely in position that it may be difficult or impossible to straighten. However, the holes at the top and bottom of the electric box are elongated to make allowance for such conditions.

Mount the receptacle so it is vertical. The electric box housing the outlet cannot be seen so that doesn't really matter, but the receptacle itself is visible and should be straight, not for electrical reasons but for the sake of appearance. Figure 4-12 shows how a receptacle can be correctly mounted even though its electric box is at an angle.

Figure 4-12 Even if the electric box is at an angle, the receptacle should be straight.

Sometimes an electric box will be pushed deeply into a wall, making it difficult for the receptacle to be flush with the wall. In that case use flat washers behind the receptacle mounting screws (Figure 4-13). A better approach is to reposition the box.

If the electric box is too deep and it is to be used for a switch, the result will be that the toggle of the switch will barely extend from its covering wall plate. It

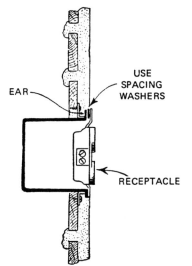

EAR

USE SPACING WASHERS

RECEPTACLE

Figure 4-13 If the electric box is too deeply recessed, bring it forward with spacing washers. This drawing shows an installation in an older home using sheetrock.

will not only look bad but will make the switch difficult to operate. Again, you may use washers behind the switch-mounting screws, but it is preferable to reposition the box. Adjust the box until the switch is in its proper position. In both cases, outlet or switch, check by mounting the wall plate. The receptacle should protrude slightly in front of its wall plate, while a switch should have its toggle as far in front of the wall plate as possible.

HOW TO MODIFY A DUPLEX RECEPTACLE
INTO A SWITCH RECEPTACLE

Ordinarily, electrical appliances are supplied with their own switches, but this isn't always true. A switch can be incorporated in the power cord connecting the appliance to a receptacle, but an alternative approach would be to modify the receptacle from a duplex to a single, as indicated in Figure 4-14. In this example the single receptacle is always live and can be used for powering an electrical device. The SPST switch is connected in series with the black lead.

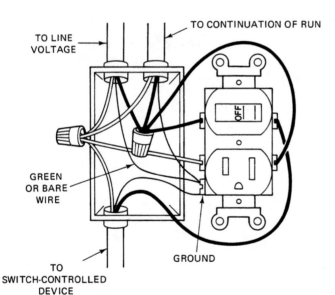

TO LINE VOLTAGE

TO CONTINUATION OF RUN

GREEN OR BARE WIRE

TO SWITCH-CONTROLLED DEVICE

GROUND

Figure 4-14 Switch and receptacle are electrically independent.

Power is brought in from the source by a black and white lead. The white lead is connected to the left side of the receptacle. Since this is a middle of the run arrangement, the white lead continues on to other receptacles. Only two wire nuts are needed for making the connections. The ground, a bare wire, accompanies both input and output power cables, is connected to the ground terminal of the switch/receptacle, and from that point also continues on to the controlled electrical fixture.

HOW TO INSTALL A SWITCH CONTROLLED RECEPTACLE

Receptacles are usually wired to operate live, but they can be wired so that the receptacle is switch controlled as shown in Figure 4-15. Note that the switch is wired in series with the receptacle and must be in its on position for the receptacle to be powered. The switch also controls any electrical device connected to the power line shown exiting at the lower left in the drawing. However, the power line that forms the rest of the run (top right) is independent of the switch.

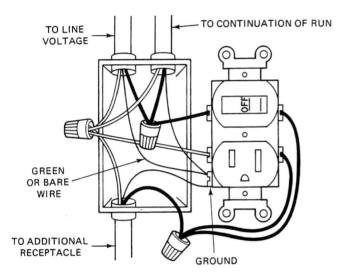

TO LINE VOLTAGE

TO CONTINUATION OF RUN

GREEN OR BARE WIRE

TO ADDITIONAL RECEPTACLE

GROUND

Figure 4-15   Switch-controlled receptacle.

HOW TO PREPARE ROMEX CABLE FOR AN ELECTRIC BOX

Some preparatory cable work is necessary before inserting Romex cable into an electric box. Put the cable on a sturdy flat surface and, using a utility knife or a single-edge razor blade, cut through the outer insulation covering of the cable and then remove it. Start this initial work about 6 inches from the end of the cable. Be careful not to cut into the wires.

Three wires will now be exposed: a white, a black, and a ground wire. Using a wire stripper, a razor blade or a pair of diagonal cutters, strip about ½ inch of the insulation from the black and white wires.

Remove one of the knockouts from the electric box, but before doing so, try to estimate the shortest path of the wires to the connections to be made to the switch or receptacle. Mount the cable connector. Push the cable through it. Tighten the screw that holds the connector to the box.

Connect the bare ground wire. This will either be to a ground clip or a screw terminal. If the wires are to be fastened to a screw terminal, form them into the shape of a small circle. If they are to be inserted in a push-in connection, simply insert the wires into the appropriate spring-loaded holes. Before doing so, however, use the wire gauge that appears on the back of the switch to make sure each wire has the right amount of exposed copper.

HOW TO MOUNT ELECTRIC BOXES

Electric boxes must be mounted rigidly and securely, for they literally serve as an anchor for a number of parts. The cables that come into the box and that leave it through knockouts do not simply go in and out loosely. They are fastened to the box, and so they are mutually supportive. The box also holds a switch or a receptacle, depending entirely on the box for support. Unlike the wiring they do not help support the box. The wall plate also uses the box as a way of staying in position, but since the four sides of the plate press against the wall, they do contribute to the overall support.

There are a number of ways of fastening an electric box, and in some instances this depends on the age of the house, since that determines the way in which it was constructed.

Side Bracket Boxes

The side bracket box is either square (Figure 4-16a) or rectangular (Figure 4-16b) and has a bracket on one side or the other for mounting the box. The box can be used for a switch, receptacle or can be a junction type.

Side brackets are generally used only in new installations, since it is difficult to install the brackets once the walls are completed. The advantage of side brackets is that the box is always at the correct depth. Side brackets are also convenient for boxes positioned on beams in an attic or basement. Since these areas are usually exposed, the side brackets can be used in either new or old homes.

A side mounting plate can be parallel to the face of the box or at right angles to it, as indicated in Figure 4-16c. The right-angle mount forms one side of the box, while the parallel type is independent.

Screw or Nail Mounting

Electric boxes are available with screw or nail holes punched out in the back, assuming the presence of a wood support. Plastic boxes may be equipped with mounting nails with these parallel to the face of the box (Figure 4-16d). The box also has screw holes in front (Figure 4-16e), so it can be supported by mounting across a pair of wood cleats.

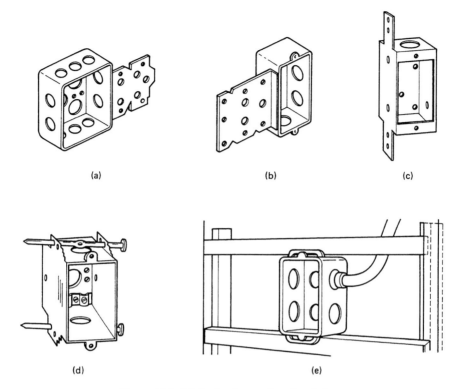

(a) (b) (c)

(d) (e)

Figure 4-16 Methods of mounting electric boxes.

The way in which an electric box is mounted also depends on the age of the house. In older houses using lath strips, the electric box is mounted in a space cut out in the lath. The box is then fastened into position by wood screws, top and bottom. Before mounting the box, decide which knockout holes to use and then punch out the knockouts accordingly. Fit the box in the space provided for it and then remove it. Make all cable connections to the box before mounting it.

Offset Hangers

Another way of mounting boxes is to use an offset hanger. The hanger may be necessary if the positioning of the box is such that it cannot be nailed or screwed into a supporting wood frame. The offset hanger comes equipped with a mounting stud that is threaded for a machine nut. Screw or nail the hanger into position (Figure 4-17).

You can get deep or shallow offset hangers. Select the type of hanger that will let the front of the box fit flush with the ceiling.

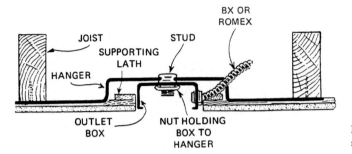

Figure 4-17 Offset hanger used to support electric box.

HOW TO INSTALL A BASEMENT ELECTRIC BOX

The easiest way to install a basement ceiling box is to run a branch power line from an existing box to a new location. If the basement is finished and has a dry-wall ceiling, the new wiring can be run through a surface wiring enclosure. An alternative approach that involves greater effort is illustrated in Figure 4-18.

Locate an existing receptacle on the floor immediately above the basement. Drill through the basement ceiling using a long-shank bit so a hole is made in the floor between studs and inside the wall behind the baseboard. Push a short length of fish wire through the hole, and you will be able to use it to pull the cable through. Since the new basement fixture will not be switch controlled, use one having a pull chain.

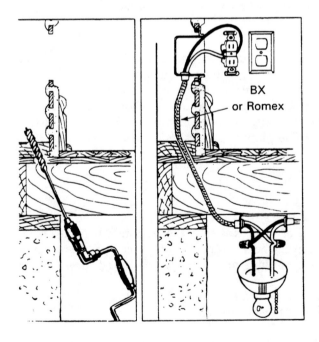

Figure 4-18 Method of installing a basement ceiling box.

To simplify the wiring, the ground wire connection is not shown. The basement lamp wires will be color coded black and white. Connect these wires, using wire nuts, to the correspondingly colored wires in the power cable from the upstairs receptacle. After the wiring has been checked, operate the pull chain to make sure the light is working. Then fasten the canopy of the new basement light to the ceiling.

Wiring of basement lights is much easier if electrical planning is done before a basement ceiling is installed.

HOW TO GANG METAL BOXES

Metal boxes can be grouped by ganging. To gang two boxes, remove the adjacent electric box walls and then fit the boxes together. They can then be joined with metal screws. This technique applies to metal boxes only and not to plastic types. The joined boxes must be covered with metal plates after the wiring is completed.

Use a ganged double-box arrangement to avoid excessive wire crowding. Ample room inside the double box makes connecting wires much easier.

Using a double box does not mean it is necessary to have an extra switch or receptacle plate. The second box can be covered with a blank metal plate. No special support is needed for the second box if the first one is kept in place by being fastened to a stud.

HOW TO ADD A WALL SWITCH TO CONTROL A CEILING LIGHT AT THE END OF A RUN

Figure 4-19 shows the wiring arrangement. In this case the end of the branch line, or feed wire, is brought into the electric box at the right. The neutral white wire will not be interrupted by the switch and is connected directly to the corresponding white wire supplied with the lamp.

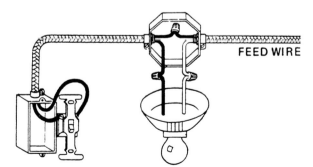

FEED WIRE

Figure 4-19 Adding a wall switch to control a ceiling light at the end of a branch.

HOW TO INSTALL A GROUND FAULT CIRCUIT INTERRUPTER

The GFCI works by comparing the amount of current flowing through the black and white wires, and it does so on a continous basis. These currents, as indicated earlier, must be equal. If there is a difference of just $\frac{1}{200}$ ampere, the GFCI will open the circuit and will do so in $\frac{1}{40}$ of a second, while some work even faster than that. The GFCI is actually an automatic overload circuit breaker.

GFCI Types

There are three variations of this component. The easiest to install is the plug-in adapter type, but this supplies the least amount of overall protection. Just plug the unit into a receptacle and the electrical device into the GFCI. This GFCI has limitations: it is intended for indoor use, and it can supply protection only for the receptacle in which it is inserted.

Another GFCI is similar to the plug-in type except that it is wired in to replace an existing receptacle. This type can be installed indoors or outdoors. The advantage of this unit over the adapter type is that it can protect either a single receptacle or a group, depending on where in a branch line it is inserted.

A GFCI can also be installed in the service panel that contains the breakers for all the power branches in the home. In this case it works both as a GFCI and as a circuit breaker. It will, however, protect only that power branch to which it is connected. More than one GFCI will be needed if several power branches are to have GFCI safety.

The GFCI is equipped with two pairs of leads, plus provision for connecting a ground lead. The GFCI component has a black and a white lead identified as *load* and another white and black lead identified as *line,* as indicated in Figure 4-20. Line means power line, so the incoming white and black power leads are connected to the GFCI's line leads. The black and white leads identified as *load* connect to wires leading to other receptacles in the power branch. When connecting at the end of a branch, cover the load leads with wire nuts for these wires will not be needed.

The front of the GFCI looks like a receptacle and is intended to work like one. It is polarized in two ways. It has provision for accepting the circular ground extension of a plug, and its slots are intended for the insertion of different-sized plug prongs. Figure 4-21 is an internal view of the wiring.

The front of the GFCI has a pair of push buttons with one used for testing, the other for resetting. These buttons are of different colors to make recognition easy.

Line Cord GFCI

Power line cords equipped with a GFCI are available for outdoor use. These cords are in lengths ranging from 6 feet to 25 feet. They have a 4 milliampere to 6 milliam-

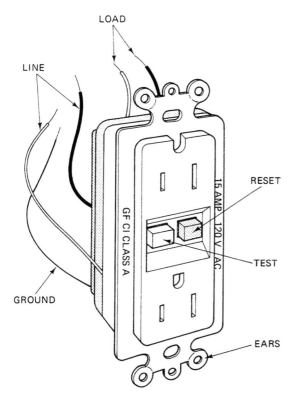

LOAD

LINE

RESET

15 AMP 120 V AC

GF CI CLASS A

TEST

GROUND

EARS

Figure 4-20 Front view of a GFCI used as a replacement for a receptacle.

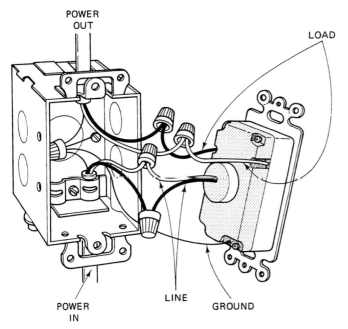

POWER OUT

LOAD

POWER IN

LINE

GROUND

Figure 4-21 Internal wiring of a GFCI.

pere trip level (0.004 to 0.006 ampere). The GFCI mechanism is waterproof while the power cord resists moisture, chemicals and sunlight. It is intended for use in boat yards, marinas, or with recreational vehicles. It supplies protection against electrical leakage from power drills, table saws, floor polishers, battery chargers, wet/dry vacuum and high pressure washers.

HOW TO ADD AN ADDITIONAL RECEPTACLE TO EXISTING WIRING

Figure 4-22 shows how to add another receptacle to an existing wiring system. All that is required is to put the two receptacles in parallel with each other.

The branch power line, the feed wire, is brought in from the upper left and is connected to the top receptacle at the left. This receptacle is connected by small metal strips to the lower receptacle. As a result these two, the upper and lower sections of this receptacle, are wired in parallel. Since the connection is done automatically by the receptacle, no wiring is required.

A white and a black wire are connected to the lower section of the receptacle and form a cable leading to the receptacle at the right. The wires are connected to that receptacle. It is wired in such a way that both the upper and lower sections are live. These two receptacles, the one at the left and the other at the right, are not switch controlled and so are always active.

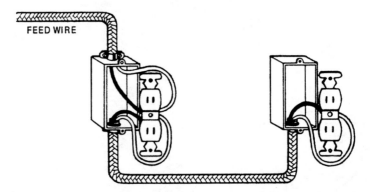

FEED WIRE

Figure 4-22 How to install a receptacle beyond an existing receptacle. The ground lead (not shown) is connected from the ground terminal of the first receptacle to the ground terminal of the second. The ground wire is brought in with the feed cable.

HOW TO FIND THE LAST RECEPTACLE IN A RUN

The word *run* is used to indicate a branch power line. The first receptacle that is wired to the beginning of the power line is the start of the run. The terminus of the power line is the end of the run. Receptacles connected between the beginning and the end are known as middle of the run. While there is only one receptacle at the beginning and just one at the end, there may be a number of receptacles referred to as middle of the run.

The end of the run receptacle has as many screw terminals as any of the others, and consists of a pair of neutral and a pair of hot screws. The screws for neutral wires are always chrome or silvery finish. Black wires are always connected to brass terminals. If push-on connections are used instead of screws, these will be labeled to indicate which wire, black or white, is to be pushed into a spring-grip hole. There is also a terminal for a ground connection, generally positioned at the lower left while you are facing the front of the receptacle. Thus, each receptacle will have five connecting points.

The end of the run receptacle can be identified by the fact that it has wires connected to only two of the four screws or only two wires using the rear-located push-ins. The two remaining screws of the receptacle can be used as the starting point for the continuation of the run. Receptacles at the beginning or the middle of a run have a pair of power-entry and a pair of power-exit wires.

HOW TO CONNECT A NEW BRANCH AT A CEILING LIGHT

New wiring can be connected at a ceiling light, assuming that the light is not controlled by a wall switch. However, the light can be operated by a pull chain.

Figure 4-23 shows the new wiring arrangement. Remove the wire nuts to make the black and white wires available for the new connections. One white wire and one black wire will now lead away from the receptacle.

There will be two ground wires, one from the power supply cable, the other leading into the new branch cable. These can both be connected to any convenient point on the metal junction box. The new run should have the same wire gauge as that used by the existing cable.

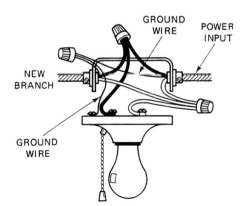

Figure 4-23 Ceiling box used for start of new branch line.

How to Control Multiple Loads

For the most part a SPST toggle switch is used to control a single load, but the same switch can operate a multiple load as well, as indicated in Figure 4-24. The three lamps are in parallel with each other but this combination is in series with the switch.

The three bulbs are operated simultaneously and all are either on or off. The sum of the currents flowing through these lamps also passes through the switch, and the switch must have an equivalent current-carrying capability. The voltage across each lamp is the same and is the amount of voltage supplied by the branch line. Ordinarily, it is 120-volts AC.

The arrangement in Figure 4-24 is often used when controlling a chandelier light comprising a number of electric light bulbs. For a fixture using a large number of lights, the current demand could be quite large and could be considered a heavy load.

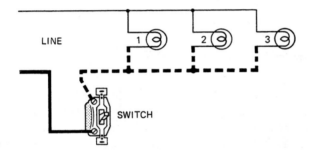

Figure 4-24   Three parallel loads in series with SPST switch.

How to Connect Switches for Individual Load Control

Quite often it may be necessary to control various loads individually, and in that case the electric circuit of Figure 4-25 can be used. Each of the switches, identified as A, B, and C, is in series with the loads 1, 2, and 3. With this arrangement, one or more of these loads can be turned on independently. Each switch must be able to carry the amount of current demanded by its own load.

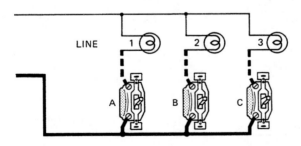

Figure 4-25   Each lamp is individually controlled.

This is also a series-parallel arrangement. Each switch is in series with its load, with this series combination in shunt with the power line.

THE SINGLE-POLE, SINGLE-THROW SWITCH

The single-pole, single-throw switch is the most commonly used switch throughout the home. It can be built into a wall, mounted on the wall's surface, made part of a line cord, or built into the frame of an appliance. It can be operated by a small plastic handle, by a metal toggle, or by a pull string.

No matter what type it may be, it is always part of the hot lead of an electrical circuit and is always in series with that lead. It is absolutely never connected from the black lead to the white lead of a branch circuit. Its function is very simple—to interrupt the flow of current from a power outlet to some electrical device.

When the switch is closed, the full current of the appliance must move through its terminals, so the current rating of the switch must be at least as high as that of the appliance it controls. When the switch is open, the full voltage of the power line is across its terminals. The fact that a switch is open does not mean it is safe to handle with bare fingers. When working on such a switch, remove its power fuse or open its circuit breaker if the switch is an in- or on-wall type. If it is part of an appliance, remove its plug from its outlet.

While there are various types of switches, they all work in essentially the same way—by interrupting the flow of current through the black lead. The white or neutral lead must remain continuous.

Connections to the SPST switch can be made from the back or the sides. Which to use depends on the way the branch wires are brought into the box that will enclose the switch. Figure 4-26 shows front and side views and supplies dimensions. There are two ways of connecting wires to the terminals of this switch. One is by wrapping the bare end of wires around a screw, then tightening it; the other is by using push-in terminals. Of the two methods the use of a screw is better but requires a little more time and work. Using a push-in terminal is easier and faster, but with this method it is possible for a wire to work its way loose.

When using a screw connection, strip the wire end and shape it clockwise in the form of a partial circle. Force it under the head of the screw so that tightening the screw helps close the wire loop. If the wire is mounted the wrong way, turning the screw will open the wire loop and may even force it away.

There is still another method of fastening a wire to a screw terminal following the procedure shown in Figure 4-27. This receptacle has a pair of holes immediately adjacent to the connecting screws. The stripped wire is inserted into a hole and, while remaining in this position, is bent around the head of the screw which can then be tightened. This unusual feature is not available on all receptacles.

The screws used on these switches are captive types, so they will open for just a few turns. Do not try to force the screw off completely, since this may damage the screw threads, making it impossible to replace the screw.

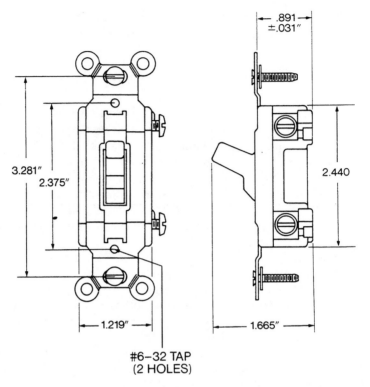

← .891 →
±.031"

3.281"
2.375"

2.440

← 1.219" →

1.665"

#6–32 TAP
(2 HOLES)

Figure 4-26 Front (left) and side (right) views of a switch showing mounting dimensions. This switch is a side screw type. (*Courtesy Hubbell Incorporated, Wiring Device Division*)

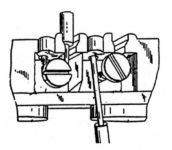

Figure 4-27 Unusual method for connecting wire to screw. (*Courtesy Hubbell Incorporated, Wiring Device Division*)

Connecting Solid and Stranded Wire

It is much easier to connect solid wire to either a screw or push-in terminal. When using stranded wire, twirl it so that it assumes the characteristics of a solid conductor. Make sure that none of the wire strands escape from under the screw head. Do not try to use stranded wire with a push-in terminal.

The toggle of all these switches is marked with the words off and on. Customarily, when it is in its on position, the controlled device receives electrical power; no power is available when the toggle is down. If the switch works in reverse, simply transpose the connections to the two screws or to the push-in terminals.

This switch, Figure 4-26, has a pair of screw-type terminals on one side. For a new installation choose a switch that will permit the easiest connections. For a replacement use a switch that duplicates the original, unless you choose to make some changes. This switch, like all the others to be described, will probably have a pair of rear push-in terminals.

Switch Data

A new switch will be accompanied by information concerning its capabilities. This will include its voltage and current ratings, the abbreviation AC, and some indication that it is UL approved. You may find the logo of the CSA (Canadian Standards Association), the Canadian equivalent of the Underwriters' Laboratories, and the words off and on inscribed on the toggle. In some older switches you may also see the abbreviations CO/ALR representing copper/aluminum. For a while aluminum was used for house wiring, a practice that has been discontinued.

How to Replace a Wall Switch

There may be two reasons for replacing a wall switch. It may have become defective, or it may be too noisy. As a start remove the fuse controlling the branch line to the switch, or open the circuit breaker. To make absolutely sure, try the switch to see if it will control its light or appliance.

Remove the two screws holding the faceplate in position, and then remove the face plate. If it is stuck to the wall, possibly because of paint, cut it loose with the point of a short knife.

You will find two screws, one above the toggle, the other below it. Remove these. You should now be able to pull the switch out of its electrical box. Take it out far enough to supply access to its connection screws or push-in terminals. Examine the way the connections are made, for they indicate how the replacement should be wired.

Turn the screws counterclockwise to release the connecting wires. Although the screws are captive types, force them out by turning the screwdriver. If the wires have a push-in connection, the old switch may be able to release these in two ways. It may have a slot for a small screwdriver or a small hole for the insertion of a short length of copper wire.

You should now have the old switch out of its electrical box, and you can put the new switch in its place. Hold the switch with the toggle in the down position with the word "OFF" showing at the top. Connect the wires to the screws or use the push-in connection. Now put the switch back in its electrical box.

If you are using a side-wired switch, you may be concerned about their closeness to the metal walls of the electrical box. Adjust the two holding screws above and below the toggle to supply the switch with a little more clearance. A side-wired switch may have a body that is somewhat thinner than the other types to ensure adequate space between the screw heads and the box. If the box is a plastic type, this precaution is not necessary.

Before replacing the face plate, put the circuit breaker back to its on position or replace the fuse. Operate the switch to make sure it works properly. Replace the face plate using its two holding screws.

HOW TO TEST A SPST SWITCH

Just because a switch-controlled electrical device or light bulb cannot be operated by a switch, do not immediately assume the switch is defective. First try replacing the bulb or device. If this does not work, check the fuse or circuit breaker. While it is possible that there is a break in the branch power line leading to the switch, it is much more likely that there is a poor connection at the switch, especially if the connections are via a push-in terminal.

Prior to removing a switch from its electrical box, turn off its power source by removing its branch line fuse or opening its circuit breaker. Just because the switch cannot control a light bulb or other load does not mean there is no power at the switch.

To remove the switch, unscrew the single centrally positioned screw holding the face plate in position. This will expose the switch.

The switch is fastened to its electrical box by a pair of screws, one at the top center and the other at the bottom center. Remove both of these screws, then pull the switch out of its box.

Replace the fuse or turn on the circuit breaker. Using a neon light line voltage tester, check to make sure that the switch is receiving power. Touch one lead of the tester to any metal portion of the electrical box and the other lead to one of the switch's connecting terminals. Then repeat this test, touching the other terminal. Try this test with the switch handle in the down position, then in the up position. You may find it easier to do these tests with one of the voltage tester's leads inserted in any convenient screw hole in the electrical box.

If the bulb of the voltage tester lights during these tests, power is being delivered to the switch. If the switch has power but is unable to turn the tester light on and off, the switch is defective.

If the connections are push-in types, turn off the electrical power and remove the connecting leads. Turn the power back on and test each lead. One of the test leads should be in contact with the electrical box. Since this is a hot test, a test with the power turned on, be careful to keep your fingers away from the bare wire ends. If the neon bulb lights during this test, you can assume the switch is defective.

The same test can be done if the box is made of plastic. With the power turned off at the fuse or breaker box, remove the push-in wires at the rear of the switch. Unscrew the wire nut joining the pair of white wires. With the power turned back on, check for voltage between the white wire and the black. Since there are two black wires, check each in turn. If during any of these tests the neon voltage tester light illuminates, the fault is in the switch.

There is always the possibility that the fault is a defective internal connection involving either the push-in or the screws, but not both. Thus, if the push-in is defective, you might still be able to use the screw connections of the switch, or vice versa.

Obvious Switch Faults

In some instances it may not even be necessary to make an electrical test. If the toggle is very loose or if it does not click when the toggle is used (even though it always did so previously), then you can be fairly sure the switch is defective.

THE GROUNDED SWITCH

A switch is automatically grounded when it is connected to a metal electrical box with its pair of mounting screws. Some of the more modern switches are equipped with a grounding terminal. Connect a wire from the screw on this terminal to a ground screw on the metal box. If the box is plastic instead of metal, connect this wire to the ground wire accompanying the cable brought into the box.

HOW TO INSTALL ADDITIONAL ELECTRIC BOXES AND SWITCHES

The diagram in Figure 4-28 shows how to add an additional ceiling outlet. It also supplies wiring information for modifying a single switch to a double, with both switches operating independently.

The feed line brought in at the top left could be from an existing junction box. The white wire is connected to the corresponding white wire from the first lamp. It also continues on to the next electric box. Here the white wire connects to the white wire from the lamp, but this is the end of the line as far as the white wire is concerned.

The connection of the two switches to the black lead is a little more involved. Two sections of Romex cable are needed for connecting the first ceiling box to the second and from the second to the switches. But Romex consists of white and black

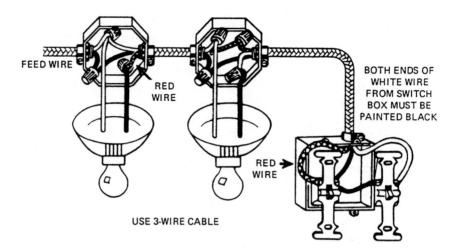

FEED WIRE

RED
WIRE

BOTH ENDS OF
WHITE WIRE
FROM SWITCH
BOX MUST BE
PAINTED BLACK

RED
WIRE

USE 3-WIRE CABLE

Figure 4-28   How to install one new ceiling box and two new switches.

wires, so in this case the white wire will be made to perform the function of a black. Put a smear of a fast-drying black paint on this white wire simply as a reminder that it is part of the hot line and isn't to be considered a neutral wire. Don't depend on memory.

The black wiring (including the one that was painted) must first go to the switches and then to the lamps. Examine the black wire in the first lamp box. It continues on to the second box and from here is connected to both switches. The switch on the right side has a white wire which returns to the second lamp. Here it connects to the black wire of this lamp. The right-hand switch now controls power to the second lamp.

The same procedure is followed with the first switch. Here we have a red wire which goes back to the first lamp connecting to the black lamp wire. Thus the first switch controls power to the first lamp.

The necessary length of red wire can be obtained from a length of Romex that consists of a white, black, and red wire. The red wire need not be painted to show that it is a hot line since red, like black, is used for this purpose.

The switches will need to be mounted in a double electric box. A pair of single boxes could be used, but the double involves less work.

HOW TO WIRE A COMBINED SWITCH AND SINGLE RECEPTACLE

The electrical circuit in Figure 4-29 consists of a switch and a single receptacle to be mounted in a double electric box. The switch is independent of the receptacle and is used to control the electrical power delivered to some component located elsewhere. The receptacle is always live, whether the switch is in its on or off position.

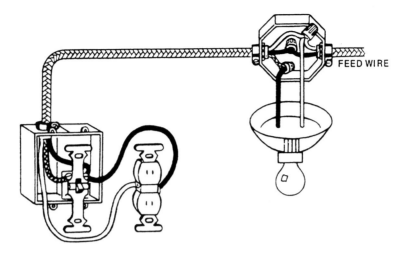

Figure 4-29   Adding a switch and a receptacle to a double electric box.

This wiring arrangement is located somewhere along the run of a branch power line. Power is brought into the box via the feed wire and makes connections to the switch and outlet.

The switch is in series with the black lead, so a black wire enters on one side of the switch and leaves on the other. The receptacle is in parallel with the power line, and a white wire connects to one side; a black coded wire to the other side.

A ground wire (not shown in the drawing) is also supplied by the source cable and continues on to the rest of the run. It also continues to the electrical device controlled by the switch. The ground wire is connected to the metal electric box and automatically grounds the receptacle and switch when they are mounted.

HOW TO WIRE A SWITCH-CONTROLLED RECEPTACLE

The single receptacle in Figure 4-30 can be controlled by the switch mounted directly above it. The unit is mounted somewhere along the branch run. The electric box receives three wires from the electrical branch: a white or neutral wire, a hot lead color coded black and a ground wire, either bare or color coded green. To make the installation, shut off the line power. Connect the wires as shown. To simplify the drawing, the ground wire has been omitted. It is connected to both electric boxes.

With this arrangement, power to the receptacle mounted below the switch must be turned on by the switch.

There are two factors involved when working with switching circuits. These are designed to work with specific switches, so selection of a circuit considered desirable must be accompanied by a switch intended for that circuit.

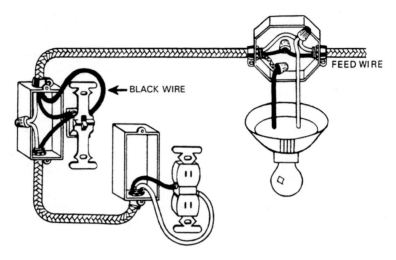

← BLACK WIRE

FEED WIRE

Figure 4-30 How to add a switch and a receptacle following an existing ceiling light.

SWITCH WIRING DIAGRAMS

Both pictorial and wiring diagrams are used in electrical work. The wiring diagram is simpler and is practically self-explanatory. With the wiring diagram it is possible to trace the circuit and to understand just how current is delivered to a load, something that can be difficuilt to do with the pictorial. While the pictorial does show the actual connections, it can sometimes be confusing. Both types, however, pictorial and wiring, are useful in learning how to make electrical connections.

Single-Pole, Single-Throw (SPST) Switching

Figure 4-31 is representative of wiring diagrams and shows the commonly used SPST switching arrangement. The two parallel lines marked neutral (white) and hot (black) that lead off to the right are connected to an AC voltage source tapped off from a junction box or a box containing a receptacle.

A variety of SPST switches are available for this circuit. One has a handle that glows when the switch is turned on, useful if the switch is to control a device that is not equipped with a pilot light. Another switch in this family also has a light which works in an opposite manner; that is, the light turns on when the switch is set to its off position. The advantage is that the light calls attention to itself in a dark room. For all three of these switches—no light, light on power on, light on power off, the circuit diagram remains the same. These switches are intended to control a single load from one location.

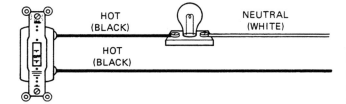

Figure 4-31 Single pole switch controls a load from one location. (*Courtesy Hubbell Incorporated, Wiring Device Division*)

Alternative SPST Circuits

Depending on the switch that is selected, alternative circuits can be used. These are shown in Figure 4-32. These switches have a glow light which functions when the load, a lamp in this case, receives current. Ground connections are not always shown, as in drawing (a). It is included in (b) and is indicated by the connection at the lower right of the switch and is represented by the ground symbol.

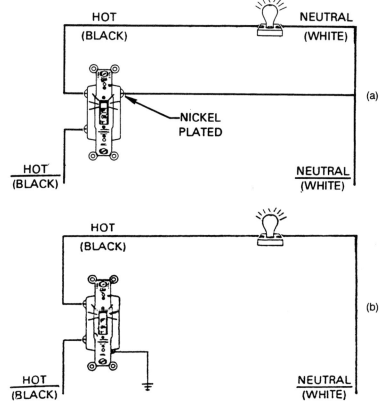

Figure 4-32 Circuit arrangements for single-pole pilot light type switches. (*Courtesy Hubbell Incorporated, Wiring Device Division*)

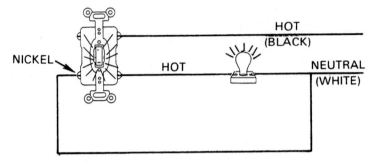

Figure 4-32 (continued)

Single-Pole Presswitch

This switch, shown in Figure 4-33, uses the same circuit as that in Figure 4-31. The difference is not in the circuit but in the switch, for it controls power by a push-to-operate switch. There are various types. One is a momentary type with the circuit normally open. The load will receive current as long as finger pressure is kept on the pushbutton. Another type works in an opposite way and will supply current to a load until the pushbutton is depressed. In both instances the circuit is the same.

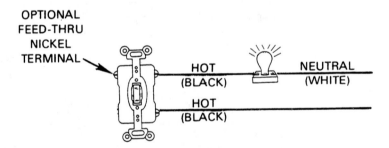

Figure 4-33 Pushbutton operated SPST switch.

Single-Pole, Double-Throw (SPDT) Switching

This circuit in Figure 4-34 controls two loads from one location. The switch is a three-position, momentary (MOM) contact type, and can supply momentary on and a non-momentary off. When the upper part of the switch is depressed, the light at the left turns on momentarily; when the bottom part of the switch is touched, the light at the right turns on, also momentarily. The center of the switch is its off position.

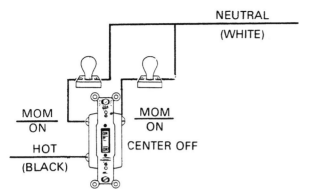

Figure 4-34 Single-pole, double-throw switch controls two loads from one location. It has a three-position momentary contact. (*Courtesy Hubbell Incorporated, Wiring Device Division*)

The circuit in Figure 4-35 also uses a single-pole, double-throw switch, but a comparison with the preceding circuit demonstrates that this description of a switch can be inadequate. Both switches perform the same function. The difference is that the one in Figure 4-35 has a three-position maintained contact; that in Figure 4-34 has a three-position momentary contact. Further, the switch in Figure 4-35 has its connections on one side only; that in Figure 4-34 has connections on both sides of the switch.

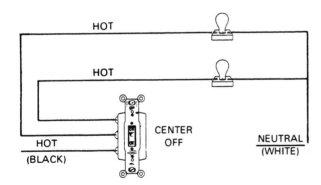

Figure 4-35 Single-pole, double-throw switch controls two loads from one location. The switch has a three-position maintained contact. (*Courtesy Hubbell Incorporated, Wiring Device Division*)

Double-Pole, Single-Throw (DPST) Switching

Figure 4-36 shows how to use a DPST switch for a single load. The switch has four connection points: two on the left and two on the right. No circuit connections are made when the switch is in its off position. When the switch is set to its on position, the hot lead is connected to the upper terminal at the left. Similarly, the neutral wire is connected to the right side of the load. The switching action puts the load directly across the power line.

There are a number of DPST switches which use the same circuit and the same connections. One of these uses a pilot light which turns on when the switch is set to

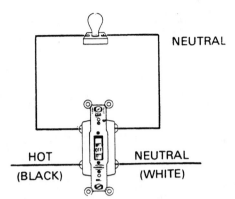

Figure 4-36 Double-pole switching for a single load. (*Courtesy Hubbell Incorporated, Wiring Device Division*)

its on position. Still another is the presswitch type with the switch using a pushbutton—push on, push off.

Double-Pole, Double-Throw (DPDT) Switching

Figure 4-37 shows two circuits that use double-pole, double-throw switches. These are three-position maintained-contact types. The three positions are off, on for one load, or on for the other. In (a) power is supplied to terminals L1 and L2. One circuit is connected to the A terminals; another to the B terminals. Either of these circuits is activated depending on the position of the switch's toggle. The two loads, shown as lights, can be positioned in different locations, so this switch and its circuit can be used to operate lights in different rooms or on different floors.

Part (b) uses the same switch and like that of (a) it is used to control two circuits. One circuit is represented by the lights at the left; the other by the lights at the right.

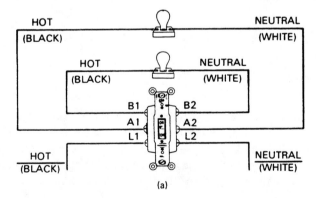

(a)

Figure 4-37 Double-pole, double-throw switch controls two circuits from one location. The switches are three-position maintained contact types. (*Courtesy Hubbell Incorporated, Wiring Device Division*)

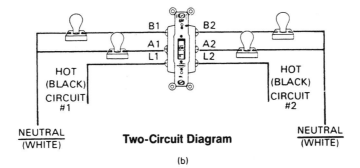

Two-Circuit Diagram

(b)

Figure 4-37 (continued)

Three-Way Switching

As indicated in Figure 4-38, a pair of three-way switches can be used to control a single load (shown here as a lamp) from two different locations. This circuit could, for example, be used to control a hall light from either downstairs or up. The switches used here are both illuminated types intended to glow when the load receives power. The two switches are identical types. The same circuit and the same connections can be used if nonilluminated switches are selected.

Figure 4-39 shows a different circuit that achieves the same result, that of controlling a single load from two locations. The change in circuitry is because dif-

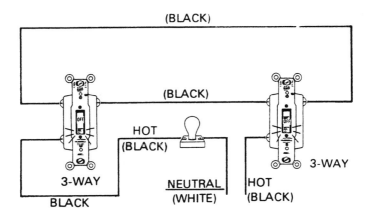

Figure 4-38 Three-way illuminated switch. The handle glows when the load is off, but the switch must have a load to make the handle glow. (*Courtesy Hubbell Incorporated, Wiring Device Division*)

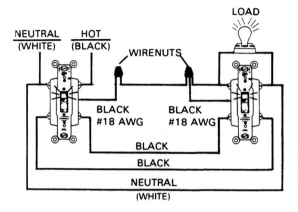

Figure 4-39 Three-way pilot light switches. Handle glows when the load is on. (*Courtesy Hubbell Incorporated, Wiring Device Division*)

ferent switches are used. These switches have glow handles that are illuminated when the load is on.

There is still another circuit variation in Figure 4-40 for controlling a single load from two locations. This indicates that not all three-way switches are alike. Further, it is inadvisable to intermix the switches. The switches in this drawing are pilot-light types. The operating switch, the one that results in load operation, is the one that glows.

In some instances a pair of circuits may look different but, upon being traced, will reveal they are identical. This is the situation in Figure 4-41. Compare it with Figure 4-38. The practical result may be that in one case a switch may be down for an operation; in the other the switch toggle may be up.

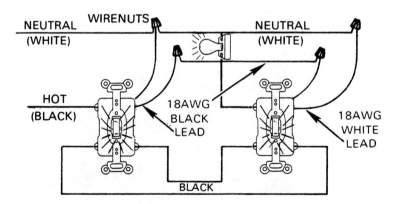

Figure 4-40 Three-way pilot light switches control one load from two locations. (*Courtesy Hubbell Incorporated, Wiring Device Division*)

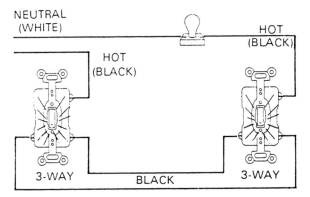

Figure 4-41 Wiring arrangement for controlling a single load from two locations.

HOW TO CONTROL A SINGLE LOAD FROM THREE LOCATIONS

It is also possible to control a single load from three different locations. It might be desirable, for example, to control a light from a basement, a first floor, and also from a second floor. This control means that the light can be turned on or off from any of these places.

The circuit is in Figure 4-42. Three switches are needed: two three-way types and a single four-way. Electrical power can come from a junction box or it can be tapped off a receptacle. The best connecting point is one whose electric box supplies space for the additional wiring. The load must not exceed the current capabilities of the branch to which it is connected.

Power is brought in by a pair of wires shown leading to the left, consisting of a neutral or white wire, and a hot or black wire. The wires that interconnect the

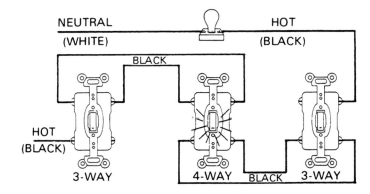

Figure 4-42 Single load controlled from three different locations. (*Courtesy Hubbell Incorporated, Wiring Device Division*)

switches are shown in the drawing as black. They can be any color other than black, but if they are the insulation should be daubed with some black paint. This is simply to indicate that these are hot wires and to act as a reminder. It would be inadvisable to depend on memory.

The four-way switch in this circuit is an illuminated type; the two three-way switches are not. While the circuit is fairly simple, the wiring may not be, especially if it is in wall and if there is some distance between the load and each of the switches.

Electric boxes must be supplied for each of the three switches.

HOW TO INSTALL A DOORBELL, BUZZER, OR CHIMES

At one time doorbells were not at all associated with home electric wiring but worked from one or more batteries. Battery replacement was a nuisance, so today's doorbell system uses a transformer connected to the AC power line.

A bell transformer is simple. It has two coils of wire not connected to each other, with both wound around an iron core. One of these coils, called the primary, is permanently connected to the 120-volt AC line (Figure 4-43). Since there is no switch in this primary circuit, current flows from the power line through the primary winding at all times. This current, known as an energizing current, is very small and the power required is negligible.

The secondary winding is connected to a pair of terminals on a bell by a switch referred to as a pushbutton. Either one of the wires from the secondary can be opened for the insertion of the pushbutton.

The switch is an SPST type, generally a pushbutton, and is mounted on the front-door jamb. See Figure 4-44 for the arrangement of the circuit. When the pushbutton is depressed, a current flows through the secondary winding of the transformer and also through the doorbell, causing it to ring. It will continue to do so as long as the pushbutton is operated.

The transformer used in the bell-ringing circuit is a stepdown type. The voltage input to the primary is the line voltage, 120 volts AC. The voltage across the secondary winding is just a few volts AC and for a bell system can be 10 volts.

There are some transformers that have split secondary windings. These transformers can be readily identified, since they have three secondary terminals. The

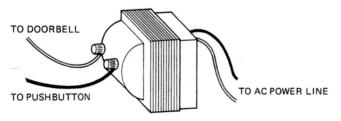

TO DOORBELL

TO PUSHBUTTON

TO AC POWER LINE

Figure 4-43 Bell ringing transformer.

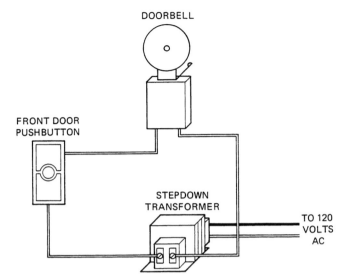

Figure 4-44 Doorbell circuit.

voltage between terminals 1 and 2 will be 10 volts; between 2 and 3 the voltage will be 16 volts; and between 1 and 3 it will be 26 volts. The reason for this arrangement is that not all doorbells require the same operating voltage. thus, a single bell transformer can be used for any of these.

The amount of current flowing through the primary winding is small but increases when the pushbutton is depressed. Since the primary is connected to the AC line, it should be handled in the same way as any other AC connection. Generally this wiring is along a basement beam to the nearest convenient AC connecting point. In some installations the transformer is mounted on the outside cover plate of the electric box to which it is connected.

The bell-ringing circuit doesn't have a separate fuse but depends on the fuse or circuit breaker of the branch to which it is connected. The primary wiring can be gauge No. 16 or No. 18 two-wire Romex. The same wire can be used for the secondary connection to the bell and the pushbutton, although Bell wire (from which it gets its name) is sometimes used. Romex is preferable, since it is much less likely to short.

Two-Terminal Doorbell

In some instances a home will be equipped with front and back doors, so the circuit in Figure 4-45 can be used. The primary connection to the AC line remains unchanged and so does one of the connections between the doorbell and the transformer. The wire between the front door pushbutton and the bell now continues on

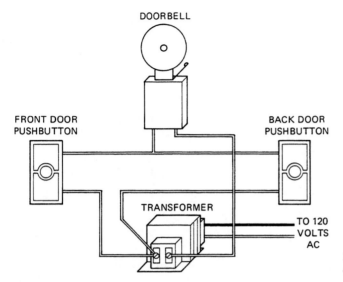

Figure 4-45 Front- and back-door control of doorbell.

to the rear door pushbutton and from that switch to the tap on the secondary winding of the transformer.

When the front door pushbutton is depressed, a current flows from the secondary winding of the transformer to the bell and through the bell back to the other tap on the transformer secondary. When the rear door button is pushed, current flows from the tap on the secondary winding through the doorbell and back to the transformer. The two pushbuttons are in parallel and depressing either of them will cause the door bell to ring.

Doorbell and Buzzer

The problem with the preceding circuit is that it may be difficult to determine whether it is the front or back door pushbutton that is being used. This can be solved by adding a buzzer to the circuit, as in Figure 4-46. The doorbell and buzzer are wired in parallel. When the back door button is pushed, the buzzer operates; the front door button operates the doorbell.

Chimes

While bells and buzzers were popular at one time and are still being used because of their lower cost, more homes are installing chimes. These may be two-tone types, as illustrated in Figure 4-47. Some chimes play tunes, and some are combinations of electric clocks and chimes. But no matter how elaborate they become, they are wired and installed in the same way as doorbells.

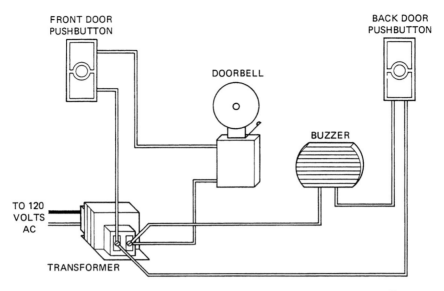

Figure 4-46 Ringing of doorbell identifies front door; sound of buzzer identifies back door.

Chimes, though, require a different operating voltage from that of a doorbell, often 16 volts. To replace a doorbell-buzzer setup, you will not need to change the wiring or pushbuttons, but you may need to change the transformer, unless a split secondary-winding type is being used.

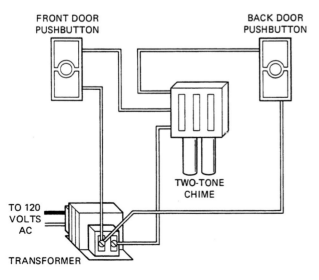

Figure 4-47 Two-tone chime identifies front and back doors.

Three-Button Chime Control

Some homes are equipped with three entry doors: front, back, and side. The wiring arrangement in Figure 4-48 is for a two-tone chime.

The door identification problem encountered with a doorbell installation also exists for chimes—identifying which pushbutton is being operated. The solution is to use a multiple-tone chime. This setup uses several tones for the front door but just a single tone for the back door.

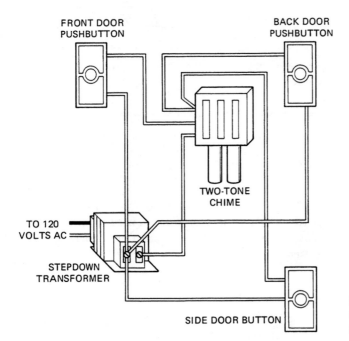

Figure 4-48   In this arrangement the front door is two tones; the back and side doors single tones only.

Troubles in the Doorbell, Buzzer or Chime System

Trouble in the doorbell, buzzer or chime system is usually due to the mechanical section, not the electrical. Trouble is most often caused by some fault in the front-door pushbutton. Some doorbell pushbuttons are equipped with a small built-in light. They are low-voltage types and are connected directly to the secondary winding of the transformer. As long as the light remains on, the primary leads and the transformer are working. Ultimately, the light will wear out and will need to be replaced, but its useful life should be about a year.

If the system uses two pushbuttons, and only one of them works, then the fault is in the inoperative pushbutton. If both pushbuttons do not work, then the fault may be in the wiring, in the bell, buzzer or chimes. Transformers have no moving parts and are the least likely suspect. In some instances the doorbell may become jammed, and in that case the doorbell, buzzer or chime will continue to produce sound. Disconnect the pushbutton and replace it. If the system is the type that supplies different sounds for front, back and side door pushbuttons, it will be much easier to locate the faulty pushbutton.

Door Opener

A door opener is a device that works the same as a doorbell or chime arrangement. When a pushbutton, often located in the kitchen, is depressed, it closes a circuit permitting a small current to run through a catcher device that is part of the front-door lock. The current releases the latch, allowing the door to be opened. At the same time the noise produced by the vibrating latch is an indication that the door is ready to be opened.

5

Indoor
Electric Lighting

FACTORS AFFECTING LIGHT

Just as many homes are inadequately wired, so too many homes use poor lighting. There is an important difference. Poor lighting doesn't always mean inadequate lighting: it can also mean excessive lighting, lighting that is too concentrated, or not soft enough. It is possible to have a large room with a strong light at one end with the rest of the room practically in darkness.

Basements and attics, particularly in older homes, are usually so poorly lighted that they are often accident traps. All of this is unfortunate because it is so unnecessary. Good lighting is easily obtainable. Further, compared to some power-hungry appliances, good lighting can be the best electrical buy. A toaster will gulp 1200 watts; a 100-watt electric-light bulb less than 10 percent of this amount.

Factors That Determine Seeing Ability

Size. The larger an object, the more it can reflect light. Drop a button on the floor, and locating it can be troublesome. It just doesn't give back enough light. But it's easy to see the dresser from which the button dropped. A larger reflecting surface means greater visibility.

Contrast. An object that is uniformly dark in color is not seen as easily as one that is made up of contrasting colors.

Reflectivity. Some colors reflect light better than others. A white wall in a room will reflect light, if there is any light available, better than a black wall.

Time. It takes time for the eyes to adapt to light. In a darkened room the eyes adjust to receive the maximum reflected light. Out in the sunshine the eyes adapt again.

Steps to Better Quality Lighting

Getting the right amount of light into the home is just one step. There is also the problem of light comfort. A 50-watt electric-light bulb for reading can be uncomfortable in one arrangement, producing eye strain, while another arrangement, also using a 50-watt light, permits seeing easily and well.

To make sure light is comfortable as well as adequate, the light in the room must be well distributed, free from glare, with no bright spots or deep shadows.

Light affects color. When selecting a particular wall paint, wall paper, or wall fabric, examine it in two ways. Get as large a sample as possible and look at it under working lighting conditions. A sample of wallpaper or fabric will have one color in the showroom and possibly another in the home because the lighting conditions are different.

When planning to paint, don't do an entire room. Try some on a section of wall and then consider it under two conditions: the first with natural light from windows, the other with artificial light from lamps. The color of the paint, its reflectivity—whether it is a gloss type such as enamel or a flat paint such as some latex types—will make a big difference in overall room illumination.

Dimmer Effect

The room lighting may be controlled by a dimmer switch. With incandescent lights, increasing the dimming action to reduce overall light level can have the effect of changing the light of the bulbs from white to yellow. This doesn't mean it's not good. Yellowish light can be soft and pleasing. That's not the question. The question is whether this is the color of light that is wanted.

Effect of White

White in a room will supply the greatest amount of reflectivity. A room with white ceilings and white or near-white walls may use lights with a lower wattage rating. This is also true if much of the furniture is either white or close to it in color. A child's room, particularly a nursery, may be done in a shade of white. But if unnceccessarily high wattage-rating bulbs are used, there may be too much glare, and in a nursery the infant can complain but not communicate.

The colors used in a room depend on the amount of light. If a room has substantial amounts of daylight, it is possible to use strong colors. Strong colors, though, may make the room look smaller; use light tints on a wall to make it appear larger.

Table 5–1 supplies the approximate percent of reflection of light by various colors.

This table supplies a basis for comparison of the different colors and the percentage of light they reflect. Darker tones, such as walnut or mahogany in woods

TABLE 5-1   APPROXIMATE REFLECTANCE VALUES

Color	Approximate percent reflection
Whites:	
Dull or flat white	75–90
Light tints:	
Cream or eggshell	79
Ivory	75
Pale pink and pale yellow	75–80
Light green, light blue, light orchid	70–75
Soft pink and light peach	69
Light beige or pale gray	70
Medium tones:	
Apricot	56–62
Pink	64
Tan, yellow-gold	55
Light grays	35–50
Medium turquoise	44
Medium light blue	42
Yellow-green	45
Old gold and pumpkin	34
Rose	29
Deep tones:	
Cocoa brown and mauve	24
Medium green and medium blue	21
Medium gray	20
Unsuitably dark colors:	
Dark brown and dark gray	10–15
Olive green	12
Dark blue, blue-green	5–10
Forest green	7
Natural wood tones:	
Birch and beech	35–50
Light maple	25–35
Light oak	25–35
Dark oak and cherry	10–15
Black walnut and mahogany	5–15

and a deep tone such as medium blue, have low reflectivity. Such a room should have ample natural light, and it will also need more wall and ceiling lighting, plus more occasional lamps.

No matter what the room size, the safest course is to reach an average: keep furniture, ceiling and wall colors in about the 30% to 60% reflectance range. The usual routine is to have ceilings painted white, while floors are much darker. This is a good procedure, since light can reach the ceiling from a ceiling fixture and is then reflected downward. The ceiling acts as a larger reflecting surface and also diffuses the light, making it softer. Further, it helps spread light over an entire room.

There are room arrangements in which the ceiling is covered by paper or fabric. Invariably these reduce the amount of reflected light, and while this problem can be overcome by adding more lights or higher-powered bulbs, the fact remains that such rooms can look gloomy if this isn't done.

Total Light

In any room there will be two basic kinds of light: direct light from incandescent bulbs or fluorescents and ambient light reflected from walls, ceilings, furniture and lampshades. Ambient light also includes light coming in through windows or any other access to outside light. The total light in a room is a combination of direct and reflected light.

ELECTRIC LIGHTS

The are two types of electrical illumination: incandescent bulbs and fluorescent lights. Both are used in the home and are often intermixed.

INCANDESCENT BULBS

Incandescent bulbs are available in just about every conceivable style, size and shape. Some are designed to flicker, simulating the light of a candle; others may supply light in a number of different intensities. Commonly the general household bulb has a wattage range from a 4 watt night light to as much as 300-watt floodlights.

Clear Bulbs

Clear bulbs are incandescent types and are so called since the glass envelope is transparent. Not all bulbs supply the same amount of light. The clear type gives the most light for a given wattage rating. Thus, a 30-watt clear will furnish more light than a 30-watt frosted or a 30-watt white. Clear bulbs are used where maximum illumination is wanted at the lowest cost. They are often used in basements, attic areas or garages.

Clear bulbs are also used in specially designed shapes, such as torch lamps for crystal-type chandeliers, particularly those that are dimmer controlled. The advantage of a clear bulb is that it not only supplies more light but light that has more sparkle. Clear bulbs do not give diffused light.

Dirt and Illumination

Electric light bulbs get dirty, with the dirt forming an opaque surface coating. While this doesn't cut out the light completely, it does reduce the amount. A dirty 75-watt bulb might supply only the equivalent of a clean 50-watt unit, but it still uses 75 watts of elecrical power. The first step in getting more illumination without using higher-powered bulbs is to clean those that are in use.

Wattage Rating of Bulbs

The wattage rating of a bulb is indicated in two ways. It will be on the packing box holding the bulb and should also be on the glass end of the bulb. The wattage marking on the bulb is often difficult to see. Breathing on its end and then rubbing it gently with a small cloth can be helpful.

Frosted and White-Finish Bulbs

Bulbs of this kind diffuse the light and help avoid excessively bright spots. For a given wattage rating, they do not yield as much direct light as clear bulbs.

Three-Way Bulbs

These bulbs have two filaments. They are called three-way because each filament can be operated separately or together, supplying three different levels of illumination. Thus one light will be 30 watts, another 70 watts, and the combination will be 100 watts.

Three-way lamps use either medium or mogul (large) bases. The mogul size is used for larger levels of light. Table 5-2 lists the various types of three-way bulbs and their uses.

TABLE 5-2   SIZES AND USES FOR THREE-WAY BULBS

Socket and wattage	Description	Where to use
Medium:		
30/70/100	Inside frost or white.	Dressing table or dresser lamps, decorative lamps, small pin up lamps.
50/100/150	Inside frost or white.	End table or small floor and swing-arm lamps.
50/100/150	White or indirect bulb with "built-in" diffusing bowl (R-40).	End table lamps and floor lamps with large, wide harps.
50/200/250	White or frosted bulb	End table or small floor and swing-arm lamps, study lamps with diffusing bowls.
Mogul (large):		
50/100/150	Inside frost	Small floor and swing-arm lamps and torcheres.
100/200/300	White or frosted bulb.	Table and floor lamps, torcheres.

GUIDE TO INCANDESCENT LIGHT BULBS

It is impossible to give absolute guidelines for electric light bulbs. A room that will seem adequately illuminated for one person may appear dim to another. Nevertheless, Table 5-3 does carry recommendations and gives suggestions on selecting incandescent bulbs for different areas in the home.

TABLE 5-3   SELECTION GUIDE FOR INCANDESCENT BULBS

Activity	Minimum recommended wattage[1]
Reading, writing, sewing:	
Occasional periods	150
Prolonged periods	200 or 300
Grooming:	
Bathroom mirror:	
1 fixture each side of mirror	1-75 or 2-40's
1 cup-type fixture over mirror	100
1 fixture over mirror	150
Bathroom ceiling fixture	150
Vanity table lamps, in pairs (person seated)	100 each
Dresser lamps, in pairs (person standing)	150 each
Kitchen work:	
Ceiling fixture (2 or more in a large area)	150 or 200
Fixture over sink	150
Fixture for eating area (separate from workspace)	150
Shopwork:	
Fixture for workbench (2 or more for long bench)	150

[1]White bulbs preferred.

Tinted Bulbs

Electric light bulbs are available that glow with all sorts of colors. Some are intended for decorative effects; others have practical uses. Red bulbs are used to supply light in photographic darkrooms and above exit doors. Yellow bulbs simulate candles, and lamps of other colors give a particular kind of atmosphere. The color is produced by a tinted silica coating inside the bulb. This reduces light output, so such bulbs supply less light than clear or frosted types.

SILVER-BOWL BULBS

These have a silver coating applied inside or outside the rounded end. The purpose of the coating is to work as a reflector, directing light upward toward the ceiling or toward another reflector.

Spotlight and floodlight bulbs belong to the silver-bowl family. Spotlights are used to direct a narrow beam of light on some object, while floodlights are used to cover a much wider area.

PAR Bulbs

PAR is an abbreviation for parabolic and describes the shape of certain types of bulbs designed for outdoor use. These bulbs are rain and snow resistant.

Decorative Bulbs

These are available in a wide variety of shapes, including globe, flame, cone, mushroom, and tubular. In some types the enclosing glass is designed with a nonsmooth finish to diffuse the light.

Heat Lamps

Not all lamps are intended for illumination only. Some in-home appliances use lamps yielding infrared heat to help soothe sore muscles and to help the skin absorb cosmetic creams. Heat lamps are available in different sizes and shapes and are not interchangeable. When buying a replacement, take along the part number of the original heat lamp and the name and model number of the original appliance.

Sun Lamps

A sun lamp is a high-pressure mercury lamp and may be made of special quartz to filter harmful ultraviolet rays. Such lamps are intended to supply artificial sunlight. These are also available in various sizes: some are tubular and others look like enlarged, clear electric light bulbs but contain a small pool of mercury that vaporizes when the bulb is turned on. Again, for replacement you will need to know the part number of the bulb and the model number of the appliance.

Some appliances are dual purpose units, containing both heat lamps and sun lamps.

FLUORESCENT TUBES

The incandescent bulb delivers light only when its filament is heated. The light is actually just a byproduct, for about 90 percent of the electrical energy delivered to an incandescent bulb turns to heat. The filament of the bulb must be made hot enough to glow.

The fluorescent tube, so called because of its shape, is a much more efficient light-delivering device and is available in various lengths, commonly from 15 inches to 60 inches, although longer tubes are made for commercial use. You can also get fluorescents in circle form. The circular type are commonly used in kitchens.

At each end of the fluorescent there is a metal cap equipped with a pair of terminals or prongs. These connect to oxide-coated filaments inside the tube. The tube may be filled with an inert gas such as argon, and, in addition, the bulb may contain a drop of mercury. The inside surface of the tube is coated with a fluorescent phosphor. This is somewhat like the phosphor covering the inside front surface of television picture tubes.

Basically, a fluorescent tube is an arc lamp. When the tube is turned on, an electric spark or current flows from the filament at one end to the filament at the other. This arc supplies ultraviolet light. If the tube were a clear type, you would see a faint glow having a purple color. The ultraviolet light strikes the inner coating of phosphorescent material, causing it to fluoresce, accounting for the name, fluorescent tube.

The two filaments inside the fluorescent work quite unlike the filaments inside incandescent tubes, for they remain on only for a second or two when the tube is first turned on. After their momentary current flow, the filaments have no further current flowing through them.

Fluorescent Efficiency

Fluorescent lamps are efficient compared with incandescent bulbs. A 40-watt fluorescent lamp produces approximately 2800 lumens. Divide this number by 40, and the result is 70 lumens per watt. The lumen is a unit of light, and the more lumens obtainable for each watt of electrical power, the more efficient the lamp. A 40-watt fluorescent produces about six times as much light per watt as a comparable 40-watt incandescent lamp.

Fluorescent Bulbs

Fluorescent lights are usually available in tubular form. Less often they can be had as bulbs and in that case can be used directly to replace incandescent types. Rated at 15 watts, they can be used to replace 60-watt incandescent types without making any electrical changes. The fluorescent bulb has a life expectancy of about six years and is designed to outperform, outlast and outshine ordinary incandescent bulbs.

The fluorescent bulb can be mounted in any position. However, it cannot be used in an incandescent dimming circuit and the bulb life will be reduced when it is used in an enclosed fixture. The price of the bulb is higher than that of an incandescent type.

INCANDESCENT VS. FLUORESCENT LIGHTING

Incandescent and fluorescent lights do not compete but supplement each other. Some fixtures have a combination of incandescent and fluorescent bulbs. Fluorescent lights are available as "cool" and "warm," so a mixture of these may produce a desired kind of light.

Table 5-4 lists the operating characteristics of incandescent bulbs and fluorescent tubes.

TABLE 5-4 PROS AND CONS OF LIGHT SOURCES

Incandescent bulbs	Fluorescent tubes
Can be concentrated over a limited area or spread over a wide area.	Provide more diffused lighting—a line of light, not a spot.
Initial cost less than fluorescent tubes.	Higher initial cost, but greater light efficiency—three to six times as much light per watt of electricity.
Designed to operate at high temperature.	Cool operating temperature. Generally about one-fifth as hot as incandescent bulbs.
Have average life of 750 to 1000 hours.	Operate five to seven times longer than incandescent bulbs.
Wattages range from 15 to 300 watts.	Wattages for home use range from 14 watts (15 inches long) to 40 watts (48 inches long).
Amount of light can be increased or decreased by changing to bulbs of different wattage because most bulbs have same size base.	Cannot be replaced by higher or lower wattage tubes.
Require no ballast or starter.	Older fluorescent fixtures require ballasts, and in some cases, starters.
Do not interfere with radio reception.	May cause noise interference with radio reception within 10 feet of the tube's location.
Suitable for use in less expensive fixtures.	Adaptable to and commonly used in custom-designed installations and in surface-mounted and recessed fixtures.
Available in colors to enliven decor and accessories. Colored bulbs are 25 to 50 percent less efficient than white bulbs.	Available in many colors (plus deluxe cool white, CWX, and deluxe warm white, WWX) at much higher light output than colored incandescents.
Gain flexibility by use of three-way bulbs and multiple switch controls or dimmer controls designed for incandescent bulbs.	Gain flexibility by use of dimmer controls designed for fluorescent tubes.

TRACK LIGHTING

A track lighting system consists of an electrified channel or track to which a number of lamp holders can be attached at any point. A single electrical power source is all that is needed. The system has a large variety of lighting functions and is also used to achieve various effects, including general lighting, accent, task lighting and wall washing.

General Lighting

A centrally located track system provides an opportunity for illuminating large rooms, hallways, entrances, bath and bedrooms. Combined with accent lighting, the system provides a multitude of lighting effects. Lampholders can be aimed straight down or at different angles to brighten every corner. A lampholder using a 100-watt R25 lamp attached to a track on an eight-foot ceiling will light a circle approximately 10 feet in diameter. A lamp with an R number is one that has an internal reflectorized coating.

Accent Lighting

You can use adjustable lampholders to direct the light onto walls to highlight art and decorative accessories. Accent lighting can make rooms appear larger by adding contrasts and shadows. It brings out textures and brightens colors. For accent and task lighting try to establish a pleasing relationship between the general- and accent-illumination levels. The accent illumination level should be three to six times brighter than the general illumination level. Contrast levels lower than three to one are generally ineffective, while contrast levels greater than six to one can become unpleasant.

Track Lighting

You can use track lighting to direct light exactly where needed without creating eye-tiring glare. Concentrate the light in a small area for work, reading, working on hobbies.

Wall Washing

Vertical surfaces can be flooded with overall light, called wall washing (Figure 5-1). This adds to the drama and spatial aspects of the interior. Another method, scalloping, draws the eye to a particular area and adds architectural interest where such detail may be lacking.

Nonuniform wall washing or rendering is achieved by aiming the lampholder at an acute angle to the vertical surface. This technique will provide texturing and a pleasing gradation of light to dark.

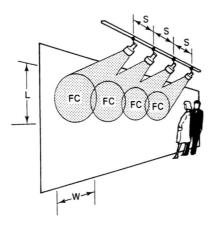

Figure 5-1 Wall washing. See Table 5-5. (*Courtesy Edison Lighting*)

To achieve even, overall illumination on a wall, use the guidelines shown in the lamp performance data chart in Table 5-5. The type and wattage of the lamps used will determine the exact placement as well as the brightness and size of the light beam on the wall.

TABLE 5-5 LAMP PERFORMANCE DATA AT 60° AIMING ANGLE

Lamps	D	FC*	L	W	S
watts/type/ size/beam pattern	Track distance from wall	Footcandles measured at beam center	Height of beam	Width of beam	Lampholder spacing
30WR20FL	2'	16	5'	3'	3'
	3'	7	7'	4'	4'
40WR16	2'	16	4'	4'	3'
	3'	7	5'	6'	5'
50WR20FL	2'	32	5'	3'	3'
	3'	14	7'	5'	4'
60WA19	2'	15	8'	8'	2'
	3'	7	10'	10'	3'
75WR30FL	2'	43	4'	5'	4'
	3'	19	6'	7'	6'
75WR30SP	2'	69	5'	2'	2'
	3'	31	7'	4'	3'
100WR25FL	2'	48	4'	3'	2'
	3'	21	6'	4'	4'

*The intensity of light is measured in foot candles (FC). A lamp, rated at 100 candle power at a distance of 1 foot supplies a light intensity of 100 FC. The intensity varies as the square of the distance. If the distance is increased from 1 to 2 feet, 2 squared = 4. In this example, the intensity of 100 FC divided by 4 = 25 FC. Light intensity in FC can be measured with an FC meter.

Aiming Angle

A major factor in lighting vertical surfaces is the correct placement of the fixture to ensure that proper aiming angles are achieved. The center of the projected beam spread (Figure 5-2) should be focused at eye level or approximately 65 inches from the floor. To avoid glare from glossy or reflective surfaces, maintain a 60 degree aiming angle from the horizontal.

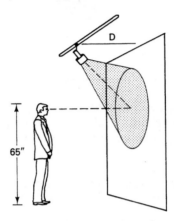

Figure 5-2   Aiming angle. See Table 5-5. (*Courtesy Edison Lighting*)

How to Connect Track Lighting

There are various ways of powering track lighting. The design in Figure 5-3a uses a cord and plug connector. This eliminates the need for permanent wiring. Just push the connector into the track. Attach the track to the wall or ceiling with toggle bolts, plug the cord into the switch and the switch into a standard wall receptacle.

Live End Connector

Figure 5-3b shows how to make the connections to a junction box. The white and black wires of the track connect to correspondingly colored wires in the ceiling junction box. Push the live end connector into the track. Mount the track with the junction box cover plate to the ceiling and then connect a switch to the live end connector.

Floating Canopy End Connector

Figures 5-3c shows the wiring. The track is connected to the junction box at any point along the track. Engage the connector to the track, connect a switch, and mount the cover plate to the junction box.

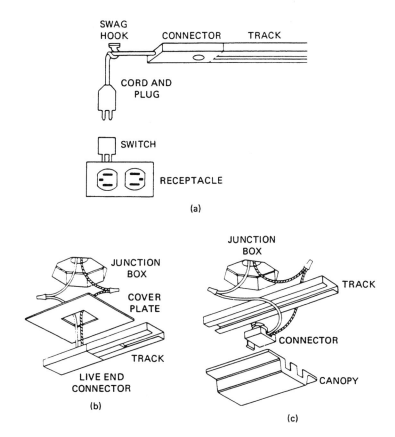

Figure 5-3 (a) Cord and plug connector; (b) live end connector, (c) floating canopy and connector. (*Courtesy Edison Lighting*)

DIMMERS

Unless otherwise controlled, an incandescent or a fluorescent lamp will supply the maximum amount of light for which it is designed. With the help of a dimmer, that light can be adjusted in smooth stages from maximum to completely off. Dimmers are generally mounted in standard, single-gang outlet boxes and are often used to replace the prior occupant of that box, a single-pole, single-throw switch. Like the switch, the dimmer uses a wall plate.

Dimmer Types

Dimmers are designed for either incandescent or fluorescent lights and are available in several arrangements, including one using a rotary control with a turn-on switch (Figure 5-4). The turn-on point supplies low level lighting which increases as its

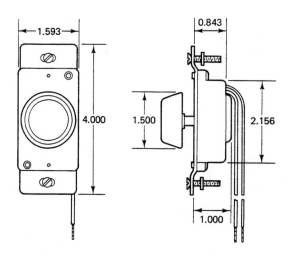

Figure 5-4 Push-on, push-off rotary dimmer. (*Courtesy Leviton Manufacturing Co., Inc.*)

control knob is turned clockwise. Another closely related type uses a push-in, push-off switch which also has rotary control of the lighting. Instead of screw-type connection terminals the dimmer is equipped with a pair of black leads.

Dimmers can also be connected into the power cords that are used with table or floor lamps. The amount of light is controlled by a wheel built into the housing of the dimmer. The dimmer is constructed so that the white lead from the line cord feeds unbroken through the dimmer while the black lead is put in series with it. Some feed-through lamp cord dimmers supply a full range of light brightness control while others are two light-level types: low, high, off.

Still another type of dimmer looks like a toggle switch. The toggle (Figure 5-5) has a switch at the beginning of its upward movement. This turns power on to the controlled lights, but as the toggle is moved upward, more power is supplied. At the limit of the vertical travel of the toggle, the lights are at maximum brilliance. There is also a three-way toggle dimmer that makes dimming possible from two

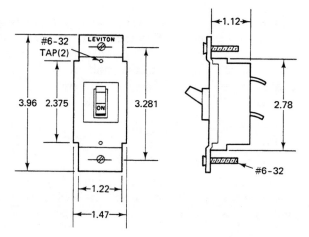

Figure 5-5 Toggle dimmer. (*Courtesy Leviton Manufacturing Co., Inc.*)

locations. A variation of the toggle dimmer consists of a slide switch that works like the toggle type.

One of the more unusual dimmers is a touch type that supplies full range control but has no moving parts. It just needs to be touched to operate. Touch it once, and it is on for maximum light. Touch and hold for full range dimming. Touch it again to turn it off. A built-in electronic memory retains the light-level setting until a readjustment is made. The unit has soft illumination at its base to make it easy to locate in the dark. A three-way type offers complete dimming and on-off control at two dimmer locations. For individual light controls the three-way dimmers can be also be mounted in a single electric box. These dimmers are not for use in four-way circuits. Flickering may occur in three-phase power systems. However, except for power supplied to heavy current demand appliances, most home electric power branches are single phase.

Dimmer Sockets

The dimmer technique can be used not only for the control of ceiling lights but for floor and table lamps as well. One of its advantages is that it eliminates the need for three-way light bulbs. The dimmer has a screw-type base which supports a screw type socket for a light bulb. The electric fixture must be a three-way type.

As indicated in Figure 5-6, the dimmer fits into an electric box in the manner of a switch. The dimmer is wired in series with the black (hot) lead.

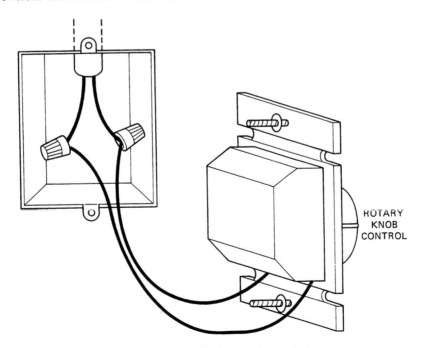

Figure 5-6 Dimmer fits into single electric box.

SOCKETS

Sockets are convenient for connecting lamps and plug-type fuses. Most sockets are screw-base types; that is, they contain threads that mate with similar threads on electric light bulbs. For small bulbs, such as those used in flashlights, the socket may be either a bayonet or a screw type.

Still another type of socket is that used by fluorescent lamps. The bulb has a pair of prongs at each end. The insertion is a snap-in action, rather than the screw-in effect of threaded sockets. Fluorescent bulbs are two-socket types; incandescent bulbs are single-socket.

Cleat Socket

A cleat socket (Figure 5-7) is a completely exposed type with the wires carrying electrical current connected to a pair of machine screws, one on each side of the socket. Since the screws are exposed, it is easy to get a shock by accidentally touching one or both screws. Cleat sockets are easy to install and can be held in place by a pair of wood screws inserted in holes at the base of the socket.

The cleat socket can also be used as a receptacle by inserting a screw-type plug into the socket. Such plugs come equipped with two or three female terminals.

These sockets require extreme care. They should be mounted so that other people, especially children, cannot reach and accidentally touch the exposed terminals. Cleat sockets are sometimes used in basements and attics and are mounted on upper cross beams. In these applications they are usually pull chain types.

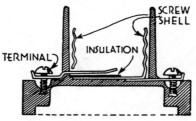

Figure 5-7 Cleat socket. Lower drawing is cross-sectional view.

Porcelain-Base Pull-Chain Sockets

The base-pull socket (Figure 5-8), resembles the cleat closely but is much safer. It receives its electrical power through a connection beneath its base, so the connecting wires are not exposed. As a further safety feature the base-pull socket is equipped with a pull chain, a switch used for turning power on and off. The base-pull socket can be mounted on a metal junction box. The box protects the wiring and also keeps it from being touched accidentally.

Figure 5-8 Porcelain pull-chain socket with receptacle. It fits any octagon electric box. The receptacle is always live. Porcelain is used to protect against rust and corrosion in damp places.

Hidden Terminal Sockets

There are various types of hidden terminal sockets but they all perform the same double purpose. The first is to conceal the connecting wires, and the other is to supply a convenient socket for a lamp or plug. Hidden terminal sockets are usually remote-switch operated—the switch isn't directly associated with the socket but is located elsewhere.

Lamp Sockets

Lamp sockets have either a pull-chain type switch or one that is knob-operated (Figure 5-9). The top part, the part opposite the threaded portion for the bulb, has various arrangements for fastening the socket in position. Some are threaded, while others are held in position by a set screw.

Figure 5-9 Lamp sockets can be pull-chain operated (left) or knob controlled (right).

Bayonet Sockets

Miniature bulbs, such as those used for flashlights, are designed for a bayonet socket (Figure 5-10). To use such a socket, insert the metal bottom portion of the bulb in the socket, push in slightly, and twist the bulb clockwise. To release the bulb and to remove it, reverse the procedure. Depress slightly and turn the bulb counterclockwise. Bulbs that use bayonet sockets are flanged types; that is, they have a small metal extension that engages a corresponding slot in the socket.

Miniature bulbs may have a screw or bayonet base (Figure 5-11).

When buying replacement lamps, it is necessary to know:

1. the voltage rating of the lamp;
2. the type of base—whether bayonet or screw type;
3. the size of the base.

Figure 5-10 Bayonet socket.   **Figure 5-11** Screw and bayonet type bulbs.

Socket Sizes

Lamp sockets are available in a number of different sizes: mogul, medium, intermediate, and candelabra, as shown in Figure 5-12. The mogul, designed for large-size bulbs, has a diameter of 1-½ inches, while the diameter of the medium socket is only 1 inch. The intermediate has a diameter of ⅝ inch while the candelabra is less than ½ inch, actually measuring about ⁷⁄₁₆ inch.

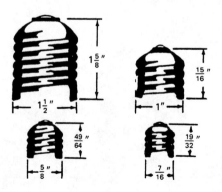

Figure 5-12 Lamp socket sizes. Mogul (top left); medium (top right); intermediate (lower left); and candelabra (lower right).

REFLECTORS AND SHADES

There are various kinds of reflectors. They resemble light shades somewhat but are different from those normally used with lamps. The purpose of a lampshade is to diffuse the light but also to concentrate it within a specific area. A reflector has a hard surface, does not spread the light, but is used to focus it.

Reflectors are sometimes used in work areas, particularly over a basement workbench. Reflectors (Figure 5-13) are often made of metal, coated green on the outside and white on the inside. The drawing shows two types, but essentially they are the same. Depending on the type used, no mechanical attachment is required and the cones and domes can be held in place by the bulbs whose light they reflect.

Figure 5-13 Cone (left) and dome (right) reflectors.

HOW TO INSTALL A CLOSET LIGHT

A light is as essential in a clothes closet as it is in any room in a home, possibly more so, since rooms may have windows to permit the entry of outside light, a benefit closets do not have.

The light for a closet can be installed in the center of the ceiling or over the door, but in either case it must not be in a position to touch any clothing. The easiest light to install is one equipped with a pull chain, eliminating the need for installing a light switch. It does mean groping for the pull chain in the dark. The best type is the one that is automatically turned on or off by the opening or closing the closet door, since forgetting to turn off a light is common. Some closet lights are battery operated, and are very easy to install, but battery replacement can be a nuisance.

The bulb size to use depends on the size of the closet, but a 75-watt type is about average. Skimping on the bulb will defeat its purpose. Fixtures are made specifically for closets, and in that case use a clear bulb. The fixture will diffuse the light.

If the closet has not previously been equipped with a light, it will be necessary to install a ceiling box and to find some way of running a power line to it. Quite often this will be through an in-wall power line connected to the nearest live receptacle. A groove can be cut in the ceiling and wall, with the power cable run through the wall to the receptacle. After the branch line has been put in the groove, hold it in position with insulated staples. Mount an electric box for a switch near the door, and then connect the black lead of the cable to it. After installing the ceiling box, bring the line through a knockout in the box and then connect the light assembly.

After checking the wiring, turn the power on by replacing the fuse or resetting the circuit breaker.

HOW TO WIRE A CEILING DROP FIXTURE

A ceiling drop fixture is any lighting fixture supported by a chain link. Before trying to get at the wiring, remove the fuse or turn off the controlling circuit breaker. Also turn off the light switch for the fixture.

Right above the lamp and up against the ceiling, there will be a small covering called a canopy, held in position by a lock nut. Loosen the nut by turning it counterclockwise, using pliers. Turn the locknut until it slides down, followed by the canopy.

Examine the wiring. There will be two white wires joined by a solderless connector (wire nut). Similarly there will be two black wires (Figure 5-14) also joined by a wire nut. One black wire and one white wire go into the fixture. The remaining white and black wires are supplied by the Romex cable that is part of the branch power line.

To replace the fixture, substitute the whire and black wires from the new fixture for the old one.

There are various ways in which the fixture is mechanically held to the ceiling. In older fixtures you will find a stud in the ceiling box. A hickey (Figure 5-15) connects this stud to the stem of the drop fixture. It sounds complicated, but it is actually simple. A stud, a hickey, and a stem are devices that fit into each other by threads cut into them. In newer fixtures the hickey is replaced by a double-threaded holder. One end of the holder screws onto the stud that is attached to the ceiling box; the other end of the holder fastens to the threaded stem of the fixture. The stem is mechanically attached to the fixture.

After joining the two wires of the lamp to the correspondingly colored wires coiled in position above the canopy, fasten solderless connectors on them, and rotate the hickey onto the stud and the stem of the drop fixture onto the hickey.

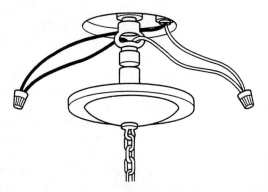

Figure 5-14   How to install a ceiling drop fixture.

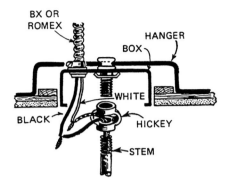

Figure 5-15   Hickey is used for joining stem to threaded stud coming out of the electric ceiling box (also called a hanger box). The hickey is found in older home ceiling drop fixtures.

There is an even easier way to learn how to do ceiling fixture replacements. After turning off the power to the fixture, lower the canopy and carefully examine the way in which the ceiling drop fixture is held in place. Disassemble these parts and observe how they fit together.

This is the mechanical side of installing a fixture. Electrically, all that is necessary is to connect the two wires of the lamp to two wires, one black and the other white, that are in the canopy. Incidentally, the two wires in the lamp aren't critical. If they are color coded, follow the coding and connect the black lead of the fixture to the black lead from the branch power line and the white lead to the white lead from the branch power line. If the wires from the light fixture aren't color coded, connect the wires without considering color coding.

Make the electrical connections as tight as possible, using gas pliers to do this. Put on the solderless connectors. Wrap them with electricians' tape (this is optional).

Check the connections before replacing the canopy. Replace the fuse or turn the circuit breaker on. Make sure the fixture has a working light bulb. Turn on the wall switch to make sure the fixture works. Put the wires back in the ceiling box and the canopy back in position.

HOW TO CONNECT A PULL-CHAIN FIXTURE

You can add a pull-chain fixture to any junction box in an attic or basement, whether that box is mounted vertically or horizontally. This type of lighting addition requires no extra cable and is simply a matter of connecting a pair of wires, as in Figure 5-16. The pull-chain fixture will have a pair of wires, one black, the other white. The fixture is often supplied with the wire ends stripped and ready for joining. Connect the black lead of the fixture to the joined black wires in the box. Similarly, connect the white wire of the fixture to the joined white wires in the box. After making the connections, cover them with a wire nut. Note that the ground wire is not shown.

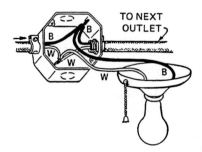

Figure 5-16 Connecting a pull-chain fixture.

A pull-chain fixture has a number of advantages. It not only supplies additional light in dark areas, but you can also unscrew the light bulb and insert a threaded plug, thus converting the fixture to a temporary convenience receptacle. The pull chain is a switch, built into the fixture.

If a junction box is not available, or if it isn't conveniently located, run a length of plastic-sheathed cable and install a new box. Use a three-wire cable (one of the wires will be a ground).

HOW TO INSTALL FLUSH-MOUNTED CEILING FIXTURES

In some cases the ceiling fixture does not have a drop chain but is mounted directly against the ceiling. The wiring of the fixture remains the same. All that changes is the method of fastening the fixture into position.

Such fixtures are mounted in kitchens, bathrooms and walk-in closets. Figure 5-17 shows that the black and white wires coming out of the ceiling electric box are connected to a pair of wires out of the fixture. The ceiling box has a fixture stud, a threaded bit of hardware that looks like the end of a small pipe.

Put the strap over the stud and then fasten it in position with a ⅜ inch locknut. The strap has a pair of holes, one at each end. These accommodate a pair of screws for holding the fixture to the strap.

What should be the order of work? Which steps should be done first? Actually, it makes little difference. If one step prevents the continuation of the others,

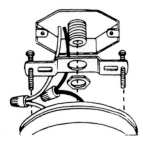

Figure 5-17 Method of attaching a fixture directly to a ceiling. The threaded center piece coming out of the electric box is the stud. Directly below it is the metal mounting bracket. The metal mounting bracket is fastened to the stud with a locknut, usually a ⅜-inch type. The two screws at the end of the metal mounting bracket are used for holding the fixture.

retreat. Undo the work and start again. If you fasten the fixture to the ceiling without joining the wires, undo the fixture to make the connections. This extreme case has happened, not only to electrical amateurs but to professional electricians as well.

The ceiling electric box may not have a stud but instead will have a pair of ears, bits of metal bent in toward the open area of the box, as in Figure 5-18. The ears are normally used to help hold a cover plate in position over the box. There will be a center hole through each ear, and these may be tapped (threaded). In that case it will be possible to fasten the metal canopy strap to the ear with a pair of machine screws. If not, put a machine screw through each ear, with the head of the screw inside the electric box. The screws will now be extending downward. Let these screws pass through the metal strap, then use a pair of nuts to fasten the strap into position.

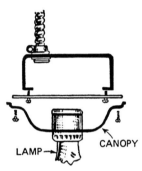

LAMP →   CANOPY

Figure 5-18   If fixture to be suspended is a small one, it can be held in place by a strap attached to the ears of the ceiling electrical box.

The drawing does not show the wiring. Simply connect the white and black wires supplied by the incoming power cable to the correspondingly colored wires of the lamp. The circuit does not show the switching arrangement.

HOW TO INSTALL OLDER-TYPE WALL FIXTURES

When preparing to paper or paint a wall that has fixtures, it may be more convenient to remove the fixture and then to replace it after the decorating work is finished. Also, you may need to remove the fixture to do some repairs, or you may want to install an older-style fixture to match others already in the room.

Electrically, old-style fixtures and more modern units are identical. You will find a pair of wires in the wall electric box, and these wires are to be connected to the correspondingly color-coded wires of the wall fixture. You may find the old wiring was taped, using a fabric type instead of vinyl. If so, remove the tape, for it has undoubtedly become too dry to maintain a firm wrap. In place of the tape, use wire nuts.

As indicated in Figure 5-19, the electric box is equipped with a centered fixture stud. A short length of threaded nipple, shown by itself in the drawing, is inserted

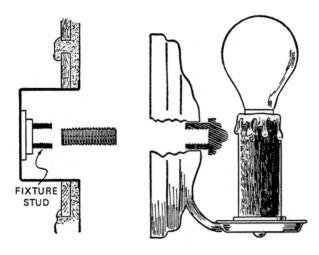

Figure 5-19 Method of installing older type wall fixture.

FIXTURE STUD

into the fixture stud and is then rotated until it makes a tight fit. The wall bracket, at the right, supports the lamp. The wall bracket has a threaded section which fits onto the nipple. When this is tightened, the lamp and its canopy will be held firmly against the wall.

When replacing the wall lamp, or repairing it, keep all the parts. It may be difficult to get replacements for these older fixtures. If that is the case, it may be necessary to substitute new fixtures for all the old-style types in a room so as to maintain uniformity of style.

At some point in the assembly there may be a problem. After tightening the cap nut, the fixture may be too loose instead of being tight against the wall. Further, it will be impossible to turn the cap nut. All this means is that the nipple extends too far into the fixture. Adjust the nipple so less of it is in the fixture. Replace the fixture on the nipple, and try to fasten the fixture into position with the cap nut once again.

If you can now make the fixture really tight, undo the cap nut to make the electrical connections. Before finally tightening the fixture, insert a light bulb and test to see if the fixture works.

HOW TO INSTALL TWO CEILING LIGHTS, WITH EACH INDEPENDENTLY CONTROLLED

The wiring arrangement in Figure 5-20 consists of two electric lights with one controlled by a wall switch and the other by a pull chain.

To understand how this circuit works, start at the lower left. A power line or feed wire is brought into the electric box, and it consists of two wires plus a ground wire. The drawing for the ground wire has been omitted.

One hot wire—the black wire—is connected to a terminal on the left side of the switch. From this point the wire continues directly to the second electric box,

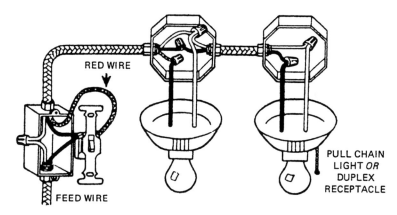

Figure 5-20 How to install two ceiling lights on the same line with independent lamp switch control.

the one at the right. It accompanies a white wire which also reaches the second light. Note that at no point is a switch included in this run. However, the second light has its own switch, a pull chain type, and that is how this light is turned on or off.

Move back to the wall switch. A wire color coded red passes through the cable to the first light and is joined to the black wire of the first lamp. The wire is color coded red, not because it is a second power line but simply to indicate that it is a hot wire. The wall switch controls power to the first lamp.

The wall switch is used to operate the first lamp, but it has no control over the second. Each lamp is independent, and each can be turned on or off either together or separately.

STRUCTURAL LIGHTING

Most in-home lighting consists of ceiling lights, desk lamps, table and floor lamps. But while these are the majority, there are variations. An electric range may be equipped with a built-in light, or it may have a hood enclosing a fluorescent. A fish tank may be illuminated or some lighting may be purely decorative.

There is also lighting to make a room more dramatic. An oil painting may have a small light mounted across the top of the frame, or a sculpture may be highlighted.

Valence Lighting

A valence is a board mounted horizontally near the top of drapes. When wired for light, it can emphasize drapery colors and accentuate the color of the fabric. Because incandescent bulbs produce a spot effect, it is better to use fluorescents. Mount the

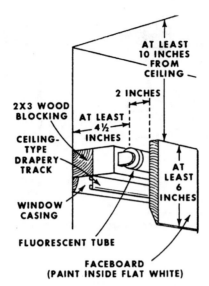

AT LEAST
10 INCHES
FROM
CEILING

2 INCHES

2X3 WOOD
BLOCKING   AT LEAST
             4½
CEILING-   INCHES
TYPE
DRAPERY
TRACK   AT
                          LEAST
                           6
                          INCHES
WINDOW
CASING

FLUORESCENT TUBE

FACEBOARD
(PAINT INSIDE FLAT WHITE)   **Figure 5-21**   Valence lighting.

tube or tubes (Figure 5-21) so they are at least 4-½ inches away from the back wall, 2 inches from the valence board, and at least 10 inches from the ceiling. Also consider using a dimmer made for fluorescent use. The switching should be independent of that for all other lights in the room.

Cornice Lighting

With this kind of lighting all the light is reflected downward, and the fluorescent fixture is closer to the ceiling. One of the advantages of cornice lighting (Figure 5-22) is that it gives an illusion of height in a room having a low ceiling. Note also

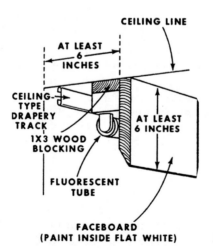

CEILING LINE

AT LEAST
6
INCHES

CEILING-
TYPE
DRAPERY
TRACK   AT LEAST
               6 INCHES
1X3 WOOD
BLOCKING

FLUORESCENT
TUBE

FACEBOARD
(PAINT INSIDE FLAT WHITE)   **Figure 5-22**   Cornice lighting.

that the fixture is mounted on the rear of the cornice faceboard. The reason for doing this is to make sure all the light will reach the entire wall surface area and to avoid shadows down at the lower end of the wall. Some experimentation may be needed to get complete light coverage of the wall.

In both valence and cornice lighting, remember that the fluorescent tubes will need replacement, so allow enough finger room. With cornice lighting the entire illuminated wall acts as a light reflector, so the amount of available light depends on the reflectivity of the wall.

Soffit

A soffit is the under part of an overhanging cornice. In structural lighting, a soffit is recessed lighting. The drawing in Figure 5-23 shows a pair of 30- to 40-watt tubular fluorescents recessed in a soffit above a kitchen sink. To make the fixture, use a faceboard that will exend at least 8 inches down from the ceiling. The installation is mounted between a pair of kitchen cabinets.

The advantage of recessed light in an above-the-sink soffit is that it supplies a warm light for easy work vision. Try using a pair of deluxe warm white types. Because the distance between the kitchen cabinets is small, you could also use incandescent bulbs, possibly rated at 75 watts and spaced about 15 inches apart, suspended from the ceiling behind the faceboard or mounted on it. Also consider such recessed lighting in other areas, such as along dressing counters in bathrooms.

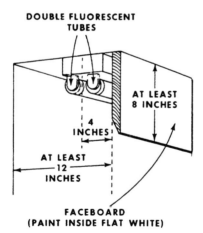

Figure 5-23   Soffit lighting.

Bracket Lighting

This lighting (Figure 5-24) is somewhat like a valence except that a valence is generally placed over a window. With a valence you would probably want draw drapes. A bracket is located over a wall, instead of windows. Placed above a sofa, it supple-

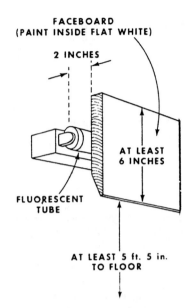

FACEBOARD
(PAINT INSIDE FLAT WHITE)

2 INCHES

AT LEAST
6 INCHES

FLUORESCENT
TUBE

AT LEAST 5 ft. 5 in.
TO FLOOR

Figure 5-24   Bracket lighting.

ments the light supplied by a pair of table lamps. Brackets are also suitable for mounting over work counters, snack bars, pictures and wall hangings.

A bracket is also acceptable for use as a bed lamp. For reading in bed, put the bottom edge of the bracket faceboard 30 inches above the top of the mattress. For general lighting mount the bracket so that its bottom edge is at least 65 inches above the floor or 55 inches for local lighting.

The faceboard of the valence, cornice, soffit or bracket affects the amount of reflected light. It is advantageous to get as much reflected light as possible, so paint the inside of the faceboard with flat white.

Luminous Ceilings

To get a soft and widely spread light, try installing a luminous ceiling. These ceilings are used for office lighting but are also suitable for some home areas, particularly kitchens. Use aluminum supports that are suspended by straps from the ceiling. The supports hold large sheets of milky plastic, sufficiently translucent to let light come through. The idea here is similar to that used for suspended ceilings, except that see-through panels are used instead of opaque ceiling tiles. With this arrangement an entire kitchen area can be uniformly lighted.

Outdoor Wiring
and Lighting

With the increased demand for indoor electrical appliances, house wiring has had to change. More branch power circuits are now being used; a greater realization that correct grounding is essential is evident; circuit breakers are replacing fuses; and lighting is more than just something to see and read by but has become a part of home decoration.

This growth in indoor electrical use has been accompanied, but possibly at a slower pace, by electricity used outdoors.

USES FOR OUTDOOR LIGHTING

There are a number of reasons why outdoor lighting is desirable. Security is improved by having lighting of general surroundings, eliminating especially dark areas or those areas having heavy shrubbery. Outdoor lighting can be used to highlight landscaping, and in so doing adds to the value of a home. It can be used to illuminate outdoor structures, such as a deck, eliminating the need to use such structures during daylight hours only. It also adds to outdoor safety by lighting walks, steps and doorways. It can minimize the possibility of lawsuits for accidental falls by visitors and non-visitors alike.

PLANNING AN OUTDOOR LIGHTING SYSTEM

Planning an outdoor lighting system is somewhat similar to planning indoor wiring prior to home construction. First, decide what outdoor electric devices will be used, including a lamp post, entry lights on either side of the front door, lights along one or more paths, emphasis lights to shine on shrubbery or a garden, floodlights along an eave on the house or a garage, lights on both sides of steps leading to the front door, and security lights which can detect the presence of a possible intruder.

Lights may also be used to decorate trees during the Christmas holidays. And while the emphasis here is on lights, consider also the desirability of having one or more outdoor receptacles for supplying power for garden tools.

Just a few of these suggestions may be used, but whichever is selected, consider also the desirability of having an outdoor electrical system that is available for future expansion. And prior to doing anything about outdoor wiring, contact your municipal authorities to determine what wiring rules to follow. It is also desirable to consult a professional electrician.

Making a Plan

You may not have been consulted about the arrangement of your indoor wiring, but you should make the decisions for the outdoor arrangement. Draw a plan indicating the devices to be used, including one or more receptacles, automatic on-off switching control for lights, building-mounted lamps, a lamp post, and the location and type of walk lights. Mark the wire size on the plan, the wire gauge and type of wire being used. Indicate all distances and the depth to which the wires will be buried and their location. Make one or more photocopies of the plan and keep one with your important papers.

Tools

Most of the tools needed for indoor wiring can also be used for an outdoor installation, including a hammer, screwdrivers, wire cutters, a wire stripper, an electric drill and bits. But you will also need a spade shovel, electrical tape, and wire nuts to accommodate the wire size that will be used.

Measurements

First decide on the location of the voltage source. This will need to be indoors and can be a junction box or the main service box. One of the factors in the selection of a junction box is its distance from the exit point of the wires to the outdoors. It is also advisable to choose a source that will be readily accessible. Measure the distance the wire will need to follow outdoors, taking into account possible curves and the burial of wire in a trench. The wire from the source to an outdoor electric device must be complete and unspliced. The alternative is to use a ground post complete with a weatherproof junction box. It is easier, better, and faster to buy enough wire to avoid this.

Overload Possibilities

Do not use a junction box simply because it is the closest to the exit point from the home, although this is an important factor. Some outdoor lighting, such as floods, can use heavy wattage bulbs. The power demand may overload an existing branch line. Add the wattage rating of all devices to be used to make sure the line can accommodate the current demand.

Wires

The wire to use should be UF (underground fused). U is an abbreviation for underground, while F represents fusing of the underground line at the service box. UF has an outer covering of heavy plastic, is available in No. 12 or No. 14 wire gauge, and may have two or three wires plus a ground lead. UF is designed for underground use without requiring conduit. However, your local electrical code may specify lead-covered UF or type TW wire and conduit. Type TW wire has a thin outer covering of thermoplastic insulation, and while this is water resistant, it should be encased in conduit for optimum protection.

Cable having a ground wire is essential. The National Electrical Code specifies the use of No. 12 wire. If the wiring is to be above ground, your municipal electrical code may call for the use of conduit. In some areas that code also indicates that outdoor wiring must be done by a professionally licensed electrician. In other localities you may do the wiring yourself, but municipal inspection will be required before power is connected.

Conduit

There are various types of conduit intended for outdoor wiring. These include aluminum and steel, both rigid types, and plastic types such as polyvinyl chloride (PVC) or high-density polyethylene. PVC is considered an above ground type; polyethylene is used for underground work. PVC reacts chemically with sunshine and so, to prevent changes in its composition, it is helpful to coat with it with outdoor latex paint. Use two coats to ensure covering the entire surface.

The inside diameter of the conduit must be wide enough to permit the use of AWG No. 12/3 UF cable (3 strands of No. 12 AWG wire). Use ¾-inch conduit.

Electrical Components

Figure 6-1 shows a weatherproof receptacle (left), a switch (center), and an electric box, with all of these designed for outdoor use. The receptacle is equipped with a weather-proof snap cover. The cover does not return automatically to its closed

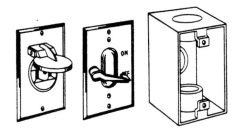

Figure 6-1 Electrical components for outdoor use. Receptacle with weatherproof snap cover, left; switches are available single and three-way, center; electric box, right.

position. Closing the receptacle is necessary to prevent the entry of rain. The receptacles are available as single and duplex types.

The switches are either single- or three-way types. The cover has a lock feature to prevent unauthorized use. Since the receptacle is always hot, remember to use the lock.

These components are available in two forms: as surface-mounted units or flush with a wall surface. The electric boxes have 4-½-inch openings. Use these outdoor devices for decorative lighting, Christmas lighting, temporary yard lighting, patio or garden lighting.

The receptacle can also supply electric power to a heating cable (Figure 6-2) to prevent ice buildup on outdoor steps or any other area where ice would be hazardous. Heating cable comes in lengths from 6 feet to 30 feet. Gutter heating cable is available in 25-, 50-, 75-, and 100-foot lengths.

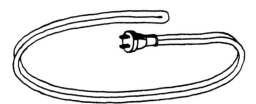

Figure 6-2 Heating cable to prevent ice buildup.

GFCI

The preferred receptacle for outdoor use is a GFCI (ground fault circuit interrupter). The unit works like a circuit breaker and can detect extremely small leakage currents. It is equipped with two pushbottons, one marked R or reset, to be used after the breaker opens, and another labeled T, an abbreviation for test. Depressing this pushbutton helps simulate the leakage of a small current.

Timers

An outdoor installation including a lamp post and wall lights on either side of the front door is switch controlled from inside the house. Using a timer instead of an ordinary switch will permit automatic switching. The timer can be set to turn the lights on and off at preselected times. A number of on-off times can also be chosen. The timer can also be used in the same way as an ordinary SPST switch.

A timer can also be inserted in series with a lamp cable after that cable has entered the home. Mount the timer where it is easily available. The wiring of the timer is the same as for any other switch. One black wire in the cable from the lamp and the other black wire from the junction box are connected to the switch terminals. The white neutral wires from the lamp and the junction box are joined so as to form a single, continuous conductor.

FLOODLIGHTS AND SPOTLIGHTS

In addition to the usual outside lights for the front of a house or garage doors, there may be occasion to use a floodlight to cover an outside area with high-intensity light. A spotlight is a form of floodlight and belongs to the same family but consists of a single bulb that focuses the light on a much smaller area.

Both floods and spots are available in clusters of two or three lights mounted on swivel holders. The swivel feature is desirable, since it permits positioning the lights for maximum effectiveness. Floodlights usually come equipped with some sort of reflecting shade and are mounted on a swivel (Figure 6-3). Spots, however, are generally not accompanied by shades.

Figure 6-3   Single unit floodlight.

Several precautions are necessary when using floods and spots. These bulbs have much higher wattage ratings than ordinary in-home electric lights and get extremely hot after working a short time. To unscrew such a bulb, make sure it has been off long enough to cool.

When using an extension cord for floods, make sure the wire has the capacity to carry the current required by the floods. If, for example, the floods require a total of 1 kilowatt, then the current (I) is equal to the power (W) divided by the voltage (E). In this example I equals W/E equals $1,000/120 = 8.3$ amperes. That's a hefty current. The connecting wire should be able to carry safely at least 10 amperes, preferably more. The usual extension cord cannot take this kind of punishment.

LIGHTS FOR OUTDOOR USE

There are a number of different kinds of lights for outdoor use, with their design depending on their function.

Mushroom Lights

These lights, illustrated in Figure 6-4, are intended for highlighting low foliage areas, borders, walkways, paths and decorative ground covers.

With this electrical fixture flowers and shrubs look much different than they do with daytime illumination. Mushroom lights are available with small and large shades, commonly 10 inch and 14 inch. The advantage of the larger shade is that it tends to conceal the light source. The smaller shade confines the light to a smaller area. Since the larger shade illuminates a broader area, such as patios or gardens, it may be necessary to use a higher wattage bulb to get the amount of light wanted.

Like most other outside lights, the mushroom concentrates its illumination in a downward pattern. To allow the light to radiate upward would be a waste of electrical power.

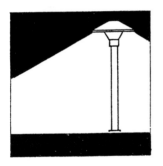 

Figure 6-4   Mushroom lights. 10 inch (left) 14 inch (right).

The Well Light

The well light, shown in Figure 6-5, has no comparable indoor light, for it is designed to be buried in the ground with all its light reflected upward. It can be placed beneath a tree or an arrangement of shrubs, highlighting the leaf structure. It can be used to call attention to a printed sign or house numbers. In the case of a well light, or for other types, several may be required to achieve a desired amount of light, or special effects. Your having selected a specific type of light does not mean you will be restricted. Not only a number of lights but a variety may be needed. If the lights are individually controlled, they may achieve a variety of effects if differ-

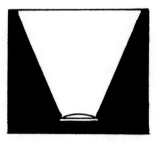

Figure 6-5   Well light.

ent lights are switched on every night. Alternatively, different lights can be turned on for the different seasons. For example, they can be used to produce dramatic results, either during or following a rainfall or a snowfall.

Entrance Lights

An entrance light, as its name implies, is to make the steps leading to a front door visible. It will also illuminate any hazards, such as a toy or any other obstruction. It also makes it easier for inserting a key into the door lock.

The entrance light (Figure 6-6) can also have an automatic turn-on and turn-off feature with the light automatically on at dusk, automatically off at sunrise. An entrance light can also be used near a sidewalk. It is also used on or near a patio for the night-time security of patio furniture and barbecue equipment.

Figure 6-6   Entrance light.

Tier Lights

A tier light (Figure 6-7) is a classically-styled fixture, suited for accenting paths, walkways and steps, and for putting attractive accent lighting along flower beds and patio borders.

The tier fixture is available in plastic or metal form and casts light downward in a soft ring of illumination while the glare of the light is shaded from the eyes. The light is projected right to the base of the fixture.

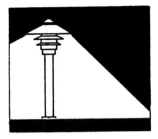

Figure 6-7   Tier light.

The shaded tier light is a variation, combining the styling of the mushroom with the tier. It features a sculptured, weather-resistant, high-impact shade.

Globe Lights

These lights (Figure 6-8) are used where diffused general illumination is wanted for a yard or garden. They are also used around the perimeter of swimming pools, hot tubs, ponds and recreational areas. A globe light provides illumination which can cover a large area without producing annoying glare.

There are two styles. One has a frosted translucent globe and the other features a shaded globe which provides a more downward-directed lighting effect.

The tier light, the shaded tier, and the mushroom light may be equipped with prismatic globes to provide a scattered form of light.

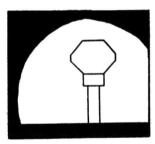

Figure 6-8   Globe light.

PROBLEMS OF OUTDOOR WIRING

In some respects outdoor wiring is easier than wiring used indoors. There are no walls to cut into; there is no required wire snaking, and there is no complexity of branch circuits to consider. However, outdoor wiring does involve the need for bringing electrical power outdoors and involves receptacles, switches, outdoor lighting, and the need for supplying electrical power to outdoor appliances. These appliances will be 120-volt types; lighting voltage can be the same or much lower, so outdoor electricity involves two different voltage levels. The higher voltage will be needed for electrically powered lawn motors, grass blowers, trimmers, edgers, and other garden equipment. Finally, outdoor electrical power will need to be fuse or circuit breaker protected.

The problem with outdoor wiring is that it is subject to rain or soil dampness. Indoor wiring works behind closed doors; outdoor wiring can involve transient pedestrians, neighbors' children, door-to-door solicitors, and animals. All of these represent potential lawsuits.

Prior to doing any outdoor wiring, contact your municipal building inspector or obtain the advice or services of an electrical contractor in your area. This does not mean you cannot do outdoor wiring yourself. It does mean being sure to follow local ordinances.

Outdoor Wiring

There are a number of factors to consider: the code for the type of wire permitted, the wire gauge, the conduit requirements, the municipal ordinance on services of a licensed electrician, the permissibility of your working under the supervision of a licensed electrician, and the use of a ground fault circuit interrupter.

Wire for outdoor use can be Type UF or TW, with conduit frequently specified. UF wire has a strong plastic outer covering and is intended for use without metal conduit. TW also has thermoplastic insulation, but it is more desirable for it to be encased in metal conduit.

It depends on whether the wire is to be put below ground or will be above it. For an underground installation use UF lead-covered cable or a dual-purpose plastic, with either having a ground wire. Again, the wire to use depends on your municipal regulations. These change from time to time, so check to make sure whether some of the newer plastic cables may be suitable. Local codes also specify wire gauge, usually AWG No. 12.

When digging a trench for underground cable, go to a depth of 1-½ to 2 feet, with the deeper one preferable. This is to avoid a possible cut-through by a tiller or spading fork.

LOW-VOLTAGE OUTDOOR LIGHTING

A low-voltage lighting system uses a number of 12-volt bulbs wired in parallel. Commonly 10 bulbs are used, while some have 12. Do not exceed the number of bulbs suggested by the manufacturer, although a smaller number is permissible. The system is flexible, for it can accommodate a programmed timer, an infrared motion detector, and a remote control. The system also uses easy-to-install ground stakes.

The power pack for the system consists of a step-down transformer, with 120-volts AC supplied to its primary winding and 12 volts to its secondary. The light bulbs are connected across the secondary, and there is no actual physical connection between the bulbs and the 120-volt AC line. The best location for the transformer is indoors, close to a junction box from which the input AC voltage can be tapped. With this position of the transformer, no provision need be made for weatherproofing it. The total wattage rating of all the bulbs that are used must not exceed the wattage rating of the transformer.

The socket assembly for each light is mounted on a spike. The lower part of the spike has a provision for accepting the connecting cable and is referred to as the fixture head. Place the cable over the metal contacts in the head. Pressing down on the cable will enable the contacts to touch the conductors. Closing the base of the head will further force the cable against the contacts. All that is then necessary is to snap the stake closed and insert it into the ground.

The wire that is used should be an underground type, but it should not have a metallic shield, as this would prevent the use of force-fit connections.

The low-voltage lighting system is fuse or circuit breaker protected at the service box. Because of the low power demand by this lighting system, it need not have its own branch. A timer can be inserted between its junction box and transformer to supply automatic lighting. Power can be supplied to the timer by the same junction box used for the system.

Because of the small amount of light supplied by each bulb, the system is not suitable for post lamp lighting. The system should also not be used for power motor-operated garden tools. Its use is restricted to low-power bulbs.

Adding More Lights

Just as indoor electrical wiring may need additional power branches, so too an outdoor revision will sometimes require changes in the lighting pattern. Since the low-voltage outdoor lighting system is stepdown power transformer operated, you may need a more powerful unit, and if the connecting cable is to be longer, you will probably need to substitute a heavier gauge.

The chart shown in Table 6-1 supplies data on the total nominal wattage of a transformer, the wire gauge to use, and its maximum length for various wattage ratings.

TABLE 6-1 TRANSFORMER SELECTION FOR OUTDOOR WIRING

Total nominal wattage of transformer	150W 16 ga. cable		200W 14 ga. cable		250W 12 ga. cable	
	max. watts	max. length (feet)	max. watts	max. length (feet)	max. watts	max. length (feet)
60 watts	60	100	60	125	60	150
88 watts	88	100	88	125	88	150
121 watts	121	100	121	125	121	150
196 watts	150	100	150	125	196	150
224 watts	150	100	200	125	250	150
330 watts	150	100	200	150	250	200

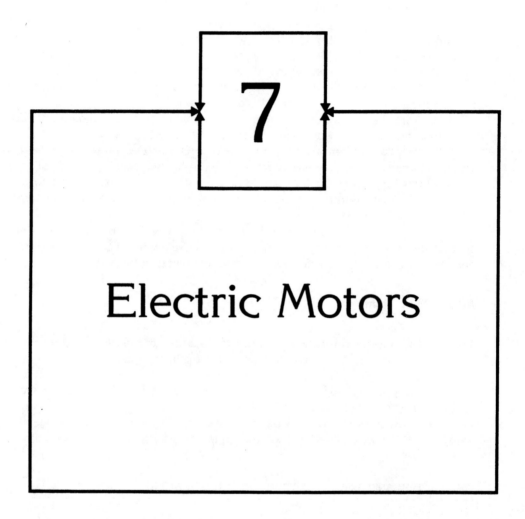

7

Electric Motors

Motors can be listed in three categories determined by the type of operating power input: AC, DC, or AC/DC. For in-home use most motors are AC or AC/DC, with this latter motor sometimes referred to as universal. Batteries are also used to supply DC power to small motors, such as those used in electric shavers, low-power vacuum cleaners, pencil sharpeners, small fans, portable clocks, and so on.

ELECTRIC MOTOR BASICS

Parts of a Motor

A transformer, such as the bell-ringing type used in the home, consists of two parts: a primary winding and a secondary winding. These consist of coils of wire wound around an iron core, both fixed in position.

A motor is like a transformer. The difference is that one of the windings in the motor, known as the armature or rotor, is allowed to rotate. The other, a fixed position winding, is referred to as the field or stator.

In a transformer the primary and secondary windings have no physical connection. Even if the secondary is wound directly on the primary winding, they are still separated by insulating material. The same is true in a motor. Generally the stator is wound so that it forms a housing surrounding the rotor. The rotor is mounted on a shaft and is fastened to it. This shaft, a rod made of steel, protrudes from the rotor and turns with it. It is this rotating shaft that supplies the torque or turning power for gears, levers, cams or other mechanical arrangements.

The Meaning of Horsepower

Horses came before machines, and in the early days of the steam engine, some comparison was needed between the amount of work done by a horse and by a machine. The term horsepower, a somewhat ridiculous yardstick, is still used in connection with motors. One horsepower, abbreviated as HP, is equivalent to 746 watts. A quarter horsepower motor is $746/4 = 186.5$ watts and a half HP motor is $746/2 = 373$ watts. You will find the HP rating of a motor on a metal plate fastened to the outside surface of the motor. Other useful bits of information on this data plate are the line voltage, the frequency (for AC motors), the power (in watts or kilowatts), and the current requirements of the motor.

The input to a motor is electrical power, measured in watts or kilowatts; the output of the motor is in horsepower. The motor, then, is a transducer, changing electrical to mechanical power.

The Motor Family Tree

Motors are often arranged by function, such as motors for sewing machines, blenders, washing machines, and so on. While this supplies information about their use, it says nothing about the type of motor. This is better done by designations such as those listed in Table 7-1.

The basic concept behind the functioning of a motor is very simple and depends on the forces of attraction and repulsion between magnets. These consist of two types: permanent magnets and electromagnets. A permanent magnet is made of metals or alloys that can be magnetized; an electromagnet is a coil of wire carrying an electric current, either DC or AC.

The armature is mounted on a shaft, supported at its ends by bearings. The rotation of the armature results in useful work that can be done by this machine.

The field coil, fixed in position, can receive its operating voltage directly. The armature does present a problem and that problem is how to supply current to a rotating coil. This is done by commutator segments made of copper and mounted in the form of a cylinder on the shaft. The copper segments are connected to the coils of the armature. A carbon brush, positioned against the commutator segments,

TABLE 7-1 TYPES OF MOTORS*

| | Alternating current | | |
Direct current	Single phase	Polyphase	Electronic
Shunt	Split-phase	Squirrel-cage rotor	Brushless, slotless, coreless
Series	Repulsion-induction	Wound rotor	Hall effect
Compound	Universal	Slip ring	Logic circuit
Split field	Capacitor	Brush shifting	Electronic control types
Differential	Series	Shaded pole	Disc stepping
Interpole			Power steppers
Universal			
Tapped field universal			

*The illustration in Figure 7-1 shows the relationships of motors.

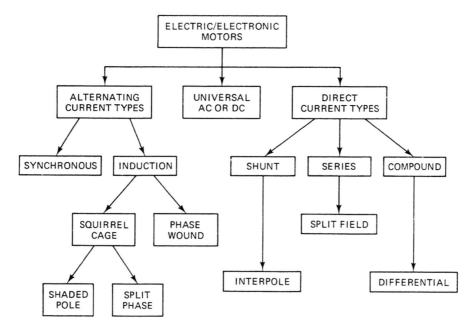

Figure 7-1   Overview of motor types.

supplies a good electrical contact without interfering with the rotation of the armature. The brushes are connected to a DC or AC voltage source. In this way an
electrical current flows from the source voltage into the brushes and then from the
brushes into the armature coils. It is this current that results in a magnetic field
around the coils.

The current flowing through the field coils also produces a magnetic field. It
is the reaction between this magnetic field and that surrounding the armature that
results in the rotation of the armature. And, when the armature turns, so does the
shaft on which it is positioned.

This is the basic way in which a motor works, and while it is simple, there are
a large number of motor types (Figure 7–1), each designed to produce certain results. Some have a large amount of turning power, known as torque; others have a
constant speed; still others can have their speed controlled easily; some are more
efficient than others; some are low-power types, while some demand a high amount
of power input.

MULTI-VOLTAGE INPUT MOTORS

Some motors are designed only for use on a 120-volt line, others for 240 volts only,
while some can have their input wiring rearranged to take advantage of both. For
high current drain devices used in the home, such as dishwashing machines, clothes

dryers, and so on, it is always better to use the higher voltage, if this is possible. But it may require installing a plug intended for 240 volt use and a receptacle to accommodate that plug. Using a higher voltage for a current-hungry motor is an advantage, since it reduces a motor's current requirements by 50 percent. Check the manufacturer's operating instructions for specific information.

The drawing at the left in Figure 7–2 shows four wires in two pairs, identified as A and B, C, and D, connected to a motor. The drawing at the left shows how these are joined for 120-volt use. When wires B and C are joined, wires A and D can be used with a 240-volt line. The motor leads may be color coded to permit an easy changeover from 120 to 240 volts.

The motor may be equipped with a data plate mounted on its frame. That plate may have a line reading "single phase AC." All homes are equipped with single phase voltage. Single phase means a single voltage. Three phase means three voltages, out of step with each other. The data plate may also read "volts 120/240, amps 20/10." This means the motor can work from either 120 or 240 volts, depending on how the connections to the motor are made and that for 120 volts the current will be 20 amperes, while for 240 volts it will be 10 amperes.

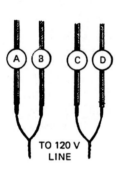
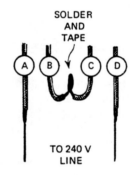

Figure 7-2 Motor connections for 120-volt line (left) and 240-volt AC line (right).

FUSES FOR MOTORS

Since the starting current of some motors is substantially higher than the running current, using an ordinary fuse will produce problems. If the starting current of a motor is 20 amperes and the running current is 3 amperes, it would seem that a logical choice would be to use a 5-ampere fuse. Yet the 20-ampere starting current will blow the fuse. For this purpose it is better to use a time-lag type. This will permit a high starting current but will open if the motor draws excessive current while working.

MOTOR SPEED CONTROLS

Motor-operated devices, such as ceiling fans and attic, kitchen, and bathroom fans, are often positioned so that operating them is inconvenient. Wall-mounted motor speed controls can be used and are designed for easy installation in a standard wall-type electric box. Some are combined with a separate dimmer, so one of the units controls light brightness and the other controls motor speed. Special wall plates are also available that indicate direction of motion of the rotary control—off, high, low. This indicates that the motor speed turns on high, with motor speed reduced as the control is turned clockwise.

Dimmers and motor speed controls are all rated for their power-handling capability, plus voltage input and line frequency. As an example, a motor speed control could have the rating 5 amps, 120 V, 60 Hz AC only. The wattage rating may be in terms of current and voltage: $5 \times 120 = 600$ watts. Also, the power rating may be stated directly.

MOTOR BRANCH CIRCUITS

If a motor requires a large amount of current, supplying that current can sometimes be a problem. There are two factors involved. One of these is the size of the wire leading from the service box to the motor. The other is the length of that wire. For most homes the distance will be less than 130 feet. However, it is necessary to divide this number by 2, since the current will flow back and forth between and motor and the service box ($130/2 = 65$ feet).

Table 7–2 shows various horsepower ratings of motors and the size of wire to use for either 120– or 240–volt service. The distances given in the table are one-way. Divide each number by 2. As an example assume a 1/4 horsepower motor located 100 feet from the service entrance and connected to a 120-volt bus. Locate 1/4 HP

TABLE 7-2   ONE-WAY DISTANCE TO MOTOR

HP	Volts	No. 14	No. 12	No. 10	No. 8	No. 6	No. 4	No. 2	No. 0
¼		140	220	350	560	890	1,400	2,300	3,600
⅓		110	170	260	420	660	1,100	1,700	2,700
½		90	140	220	350	560	890	1,400	2,200
¾	120	60	100	160	250	400	640	1,000	1,600
1			80	130	200	320	450	800	1,300
¼		560	890	1,100	2,250	3,600	5,700	9,000	
⅓		420	660	1,050	1,670	2,600	4,200	6,700	
½		350	550	880	1,400	2,200	3,500	5,600	8,900
¾		250	400	640	1,010	1,600	2,600	4,200	6,500
1	240	200	320	500	800	1,300	2,000	3,200	5,100
1½		140	220	350	560	900	1,400	2,300	3,600
2		110	170	270	430	690	1,100	1,800	2,800
3			190	310	480	860	1,200	1,900	
5			190	290	470	740	1,200		
7½				210	320	520	820		

in the first column at the left. A horizontal line separates the 120-volt from the 240-volt ratings, so in this case examine the listings above this line. The first column is for No. 14 wire. Directly below the designation No. 14 is 140, or 140 feet. Divide by 2 and the result is 70. But since the motor is 100 feet away, move to the next column. Under the heading of No. 12 is 220. Divide 220 by 2, and the result will be 110. Because the distance involved is only 100 feet, it is safe to use No. 12 wire.

The voltage drop or loss along the line connecting the motor to the fuse or breaker box will be about 3 percent. If the line voltage is 120, 3 percent of 120 is 3.6 volts. The voltage appearing at the input terminals of the motor will be 120 minus 3.6 or 116.4 volts. Wire size becomes increasingly critical for motors having a higher horsepower rating.

The Basic DC Motor

The basic DC motor, as shown in Figure 7-3, consists of a magnet supplying a north pole and a south pole surrounding an armature. The armature receives its current from a DC voltage source via a pair of carbon brushes resting on a split metallic ring referred to as a commutator. This arrangement is necessary to supply current to the rotating armature. The flow of current through the armature makes it into an electromagnet. The north and south poles shown above and below the armature can be a permanent magnet or can consist of coils through which a direct current flows. These coils are the field magnet.

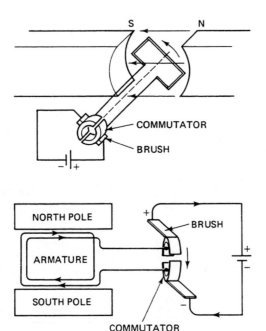

Figure 7-3 Two views of the basic structure of a DC motor.

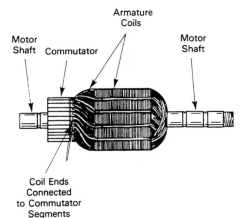

Armature
Coils

Motor
Shaft   Commutator

Motor
Shaft

Coil Ends
Connected
to Commutator
Segments

Figure 7-4   Completed armature
assembly with armature coils connected
to commutator segments.

Figure 7–4 shows the armature core, the commutator, and the shaft on which both are mounted. The armature in this illustration consists of thin iron or steel laminations. These have slots to accommodate the armature coils (Figure 7–5).

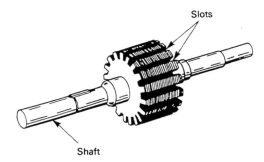

Slots

Shaft

Figure 7-5   Armature laminations
assembly on motor shaft.

THE SHUNT WOUND DC MOTOR

In this motor type the field coil and the armature are connected in parallel, with the field coil consisting of many turns of fine wire. This motor has practically constant speed even with a varying load. A variable resistor connected to the field winding can be adjusted to obtain higher speeds. Lower speeds can be obtained by putting a resistor in series with the armature. Another method of speed control is through an adjustment of the output of the voltage source.

Motor Bearings

Support for the parts of the motor are in two categories: the frame and the bearings. The frame is the outer housing and supports the field and armature coils and the

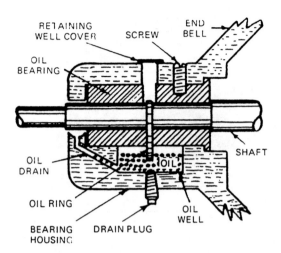

Figure 7-6 Cross-section of bearing.

motor shaft. The rotor, its windings, the commutator, and the shaft are supported by bearings (Figure 7-6). The brushes are supported in a holder and are held against the commutator by springs.

The shaft rests on end bells which support the bearings. It is essential for the correct lubricant to be used. The difference between the outside diameter of the shaft and the inside diameter of the bearings is between 0.00020 inch and 0.0010 inch, and the shaft clearance ranges from 0.00010 inch to 0.000050 inch. While the bearings are presumed to support the shaft, it rides on an extremely thin film of oil.

The type of bearing oil is extremely important and must be that recommended by the motor manufacturer. Some bearings, for example, use only pure mineral oil. The oil not only minimizes friction but prevents rust and supplies cooling. If the bearing heats too quickly, it will expand more rapidly, possibly resulting in seizing, a condition in which the shaft can no longer rotate.

BRUSHES

Brushes (Figure 7-7) are made of various substances including carbon, carbon and graphite, or a mixture of carbon and metal. Brushes carry current from an outside voltage source to the rotor.

Field Coils

The field coils (Figure 7-8) are the stator, the nonmoving part of a motor. The magnetic field of the stator can be supplied by one or more permanent magnets or by an electromagnet.

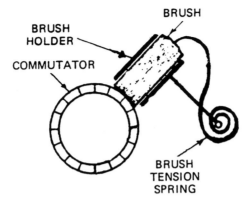

Figure 7-7 Brush, supported in a holder, makes contact with commutator segments.

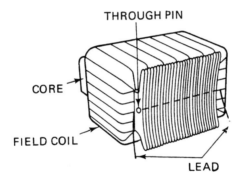

Figure 7-8 Through pin method of holding field coils in place.

The field coils are held in position against the inside portion of the motor frame (Figure 7-9). The armature is mounted on a shaft and rotates in the circular area supplied by the field coils.

The Commutator

The commutator is made of rectangular segments of hard copper, with each of these insulated from the others. The commutator segments are connected to the armature coils, and since they are also mounted on the motor shaft, rotate with the armature.

Figure 7-9 Motor frame.

THE SERIES WOUND DC MOTOR

This motor consists of a field coil wound in series with the armature. The motor is load sensitive. Its armature speed will be less with a heavy load, but as the load is reduced, the armature will run faster. It is possible for the armature to race in the absence of a load.

The series DC motor is a quick-starting type and has maximum torque at low rotational speed. In the event of a shorted armature turn, there could be excessive current flow in the shorted winding, possibly accompanied by brush sparking, with blowing of the line fuse or circuit breaker.

THE COMPOUND WOUND DC MOTOR

This motor is a combination of the series and shunt wound types and has good speed regulation. See Figure 7–10. Compound wound motors are selected for working conditions in which the machine must be started under load, but with constancy of speed not essential. The motor can supply a high starting torque.

THE SPLIT-FIELD SERIES WOUND MOTOR

When a modification is made in a motor, it is often supplied with a new name, which may or may not relate to the original motor design. The split-field motor is a series wound type with a tap at the electrical center of its field coil. The advantage of this circuit change is that a change in the direction of armature rotation can easily be obtained by using a single-pole, double-throw switch. The motor has the same characteristics as the series wound.

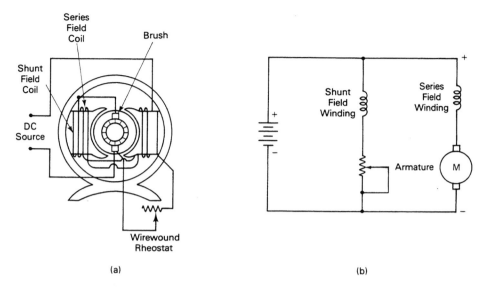

Figure 7-10 Compound wound motor. (a) pictorial; (b) schematic.

THE UNIVERSAL MOTOR

This motor is so called since its input electrical power can be either DC or AC. One of the characteristics of this motor is that it runs at very high speeds when it isn't loaded—when its shaft is turning but not driving a device, such as fan blades or the cutting blades of a food processor. The turning speed of the motor can be very high at the start but will drop to a much lower amount when the motor is put to work. In a vacuum cleaner, for example, the motor tends to race when the cleaner is lifted off a rug but decreases the moment the vacuum cleaner is moved back and forth across it.

For some universal-motor-operated devices, the load remains constant. An electric fan is one example. The fan blades are always fastened to the shaft, so the load (the blades) remains constant.

The speed of this motor can be changed by putting a resistor in series with the power line connected to it. The resistor can be a type that is tapped, permitting the selection of a few different values. In some cases the resistor is a variable unit, as in electric drills with speed selection controlled by a spring loaded trigger.

A better method for controlling speed is by using one or more taps on the field winding. The positions of the taps are selected so that more or less of the field winding is used.

The direction of rotation of the armature of this motor cannot be changed by transposing the plug, and in most instances anyway, this cannot be done. Direction reversal can be obtained, however, by reversing the direction of current flow through either the armature coil or the field windings, but not through both. The performance of the motor is not the same in both directions. Appliance manufacturers select the direction that supplies best operation of the motor.

Construction of the Universal Motor

Figure 7–11 shows the construction of a small universal motor having a range of 1/20 HP to 1/8 HP. Both the field coils and the armature coils are wound on thin sheets of steel known as laminations. The laminated armature core has slots on which the coils are wound. The armature coils and the laminated core is mounted on a shaft supported by bearings at both ends.

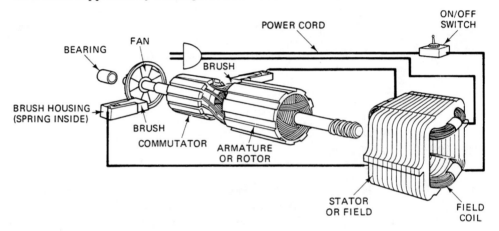

Figure 7-11   Construction of a universal motor.

Electrical power is supplied via a commutator with the segments of this component wired to the ends of the armature coils. Current is supplied to these coils by carbon brushes. In some motor assemblies the shaft is connected to a set of gears to supply more turning power but at a slower speed.

AC MOTORS

The basic difference between DC and AC motors is in the type of voltage input. The DC motor requires a DC voltage, and in the home this can be supplied by a battery for smaller motor types or by an electronic power supply that converts the AC line voltage to DC. Battery-operated motors for the home are small devices. For the most part motors for the home are AC operated.

One of the differences between DC and AC motors is the use of a commutator for DC units. One of the functions of the commutator is to change the DC into a varying DC. The DC motor doesn't use commutators exclusively, for they are also

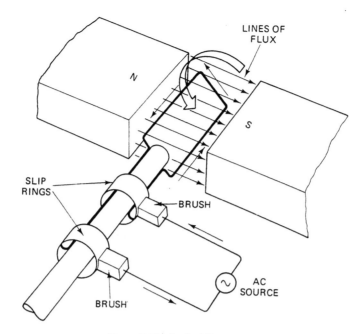

Figure 7-12   Basic AC motor.

used by a few AC motors. Usually, the AC motor is equipped with slip rings, as shown in Figure 7–12. These deliver the alternating current from the AC power line to the rotor.

There is a large variety of AC motors, much more so than DC. Some of the AC motors include the AC series, autotransformer, capacitor, single- or two- or three-phase types, squirrel cage motors, dual voltage types, synchronous, reluctance synchronous, universal, repulsion start, and so on. One of the confusing aspects is that the same motor may have several different names. Sometimes a minor change is made in a motor, and it is then supplied with a new name.

Slip Rings

Slips rings are supplied in pairs and are continuous bands of metal. They have no physical contact with each other and are insulated from the metal shaft on which they are mounted.

Brushes

Contact with the slip rings is made by brushes, and as in the case of DC motors the brushes are made of carbon or can be metallicized types.

Field

The field can be supplied by a permanent magnet or produced by a current flowing through field coils. This current is supplied directly to the field coils, and since these are not in motion they do not require the use of slip rings or a commutator.

SYNCHRONOUS MOTORS

This motor is called synchronous since its speed is synchronized to some multiple of the power-line frequency, most often 60 hz or 60 cycles per second (cps). It maintains its speed except when it is heavily overloaded, but it is regarded as a constant speed motor. Its running accuracy is evidenced by the fact that it was favored as a clock motor.

Hysteresis Synchronous Motor

This is a modification of the synchronous motor in a form known as the hysteresis synchronous. Its advantage over the synchronous motor is that it is self-starting. If, for some reason, power to a home is interrupted, this type of motor will start again as soon as power is restored.

Older electric clocks used a synchronous motor that had to be started by twirling a small metal knob on the back of the clock. Modern clocks use synchronous motors because of their accuracy, but the type that is selected is the hysteresis synchronous. Hysteresis synchronous motors are also used in phonograph turntables, not only because of their self-starting feature but also because of their ability to maintain a constant speed whether loaded or unloaded.

SPLIT-PHASE MOTORS

Washing machines generally use a motor of this type. The rotor is a squirrel-cage type (Figure 7-13), and there are two stator windings. One of these is connected to the power line at all times when the appliance is turned on; the other is used as an auxiliary winding and works from the time the motor starts until it reaches working speed.

Like a car, an electric motor needs more starting than running energy; hence, it may have two stator windings. Once the motor is running, the extra torque supplied by the second stator is no longer needed. The second stator is automatically disconnected by a centrifugal-type switch on the rotor. As the motor turns, a lever on the switch is pushed in an outward direction by the motion. When it reaches a

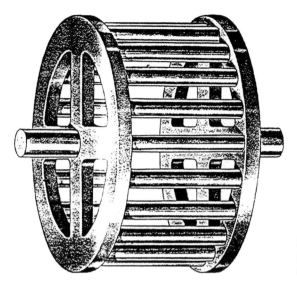

Figure 7-13   View of squirrel cage rotor showing the end pieces, the copper connecting rods and the shaft.

certain position after the rotor has achieved operating speed, it will open a switch cutting off the flow of current through the windings of the second stator.

Split-phase motors require a large starting current that can be as much as six times the running current. A 1/4 horsepower motor may demand as much as 35 amperes starting current, but this current will gradually decrease as the rotor reaches full operating speed.

A washing machine may use a split-phase motor. If it does, there are disadvantages in overloading the unit. With very heavy loads the motor draws a high current and will continue to do so until it reaches its working speed. But if it takes the motor a long time to do so, the windings may burn out. The motor winding may be designed to carry a large current, such as 35 amperes, for a reasonably short time. This large current, however, heats the coil windings of the motor, and this in turn may cause the insulation to char. The charred material will fly off as the motor rotor turns, so the adjacent turns of exposed copper may touch and short against each other. This will increase the current drain to the windings, which will get even hotter and cause still more insulation to char. The process is a cumulative one, and finally the winding takes so much current from the power line that the fuse or circuit breaker opens.

The fault is not in the motor but is due to overloading. The split-phase motor works on AC only and usually has a maximum rating of 1/3 horsepower.

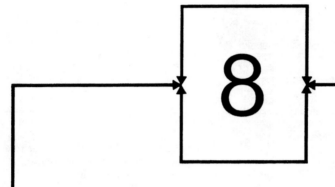

8

Maintenance
and Electrical
Troubleshooting

There are two types of electrical troubleshooting: servicing the home-wiring system and repairing the components connected to that system, such as refrigerators, air conditioning systems, electric ranges, and so on. Repairing these components, and others in this class, requires specialized expertise, plus the ability to take the component apart and the know-how to put it back together again. Even if a defective part can be found, there is still the question of being able to locate its distributor and buying it. Component parts are usually identified by parts numbers, and it is generally necessary to know these before a purchase can be made. To complicate the problem still more, each manufacturer produces a number of different models, and each of these usually has its own repair problems. Unless the repair is a minor one, it is advisable to use the services of an electrical technician supplied by the manufacturer.

When buying an electrical device, such as a refrigerator, electric stove, washer or dryer, it is best to keep the instruction manual, for this will often supply the name of repair depots, their locations and telephone numbers. Keep your purchase receipt and warranty with the instruction manual. In some instances you will find that it will be uneconomical to pay for the repairs of an appliance that has seen many years of service. In any event, if an applicance is troublesome, make a list of the problems. Does it work intermittently, create noise, generate radio and television interference, fail to do the work it is supposed to do, get unusually cold, or hot, generate smoke, produce a burning odor? When calling for service, be sure to supply the date of purchase, the name of the vendor, the model being used, the serial number, and the problem.

However, there are repairs you can make and maintenance steps you can take to keep your home electrical system in good working order. Even if that system works well, it most likely will need updating. Aside from supplying you with more electrical conveniences, updating will add to the value of your home. Further, there are many repairs or wiring installations you can do, for they are not difficult.

Many electricians work hot, that is, they do not disconnect power at the service box when working on an electrical problem. A much safer procedure is to shut off electrical power in several ways: remove the fuse controlling the branch in which the fault occurs or open the circuit breaker. Turn off any switches. Disconnect appliances from receptacles. And, even though it may seem unnecessary, test for the presence of voltage with a neon bulb tester.

MAINTENANCE CHECK LIST

Incandescent bulbs and fluorescent lamps need replacement more often than parts such as receptacles or switches. But all electrical parts require maintenance, and even bulbs and lamps work less efficiently (supplying less light) if covered with a fine layer of dust. Lamp cords can become worn, and the prongs of plugs can become spread. Instead of waiting for some part to stop working, it is better to have a regular maintenance schedule. If an electrical device does stop working, it will usually do so at a most inconvenient time.

ELECTRICAL MAINTENANCE

Switch Maintenance

Of the three commonly used electrical parts—switches, receptacles and plugs—the one most commonly subject to faulty operation and failure is the switch. Maintenance involves examination of every switch in the home electrical system with a view to anticipating possible problems. A procedure for adequate maintenance involves a series of questions.

1. Does every switch in the home have a wall plate? Is the wall plate cracked? Is it held firmly in place?
2. Is the switch loose? Does the switch move back and forth when the switch toggle is operated? With the lights in the room off and a test made in the dark, is any sparking visible?
3. When the switch is turned to its on position, is there positive action by the controlled part? If the part is a light, does it turn on at once? Does it turn off at once? Does the light flicker as the switch toggle is touched?
4. Does the cover plate of the switch feel warm to the touch?
5. If the switch is not a noiseless type, does it make a positive clicking sound when turned on and off?
6. If the switch is a glow type, does it light when the switch is on? When it is off?
7. Is the toggle recessed too deeply so that it barely extends from its wall plate?

8. Is the switch placed high enough so it is out of the reach of small children?

9. If the switch is to be replaced, what type should be substituted? A noiseless? A lighted unit? if lighted, should the light glow when the switch is on? When it is off? Do you want to replace the face plate?

Receptacle Maintenance

1. Is the receptacle firmly in position? Must you hold the face plate to keep the receptacle from moving when you insert or remove a plug?

2. Does the receptacle extend too far from the wall? Is there too large a space between the wall and the face plate of the receptacle?

3. Is the cover plate in good condition? Is it cracked? Is it being held firmly against the wall? Is the cover plate mounted at an angle?

4. Is the receptacle one that is environmentally correct? Are you using an indoor receptacle outdoors?

5. Does the receptacle hold its mating plug firmly? Does the plug have a tendency to come out too easily from the receptacle?

6. Does the cover plate feel excessively warm?

7. Is the receptacle a ground fault circuit interrupter (GFCI)? If yes, how often is it being tested? Have you set a monthly date for making this test? Is its cover plate in position? Is the GFCI out of the reach of children?

Plug Maintenance

1. Are the prongs of the plug firmly in position, or do they seem to be loose?

2. Is the plug a nonpolarized type? Are you planning to replace it with a polarized unit?

3. Does the plug feel too loose when inserted in a receptacle?

4. Do the prongs of the plug feel hot when removed from a receptacle?

5. Must you wiggle the plug to get its connected device to start working?

6. Does the plug-connected device work intermittently? Must you keep reinserting its plug to get the device to work?

7. Does the plug appear to be damaged? Has it been stepped on?

8. Is the plug one that must be pulled out regularly? Are all its users aware that a plug must be removed by its body and not by its line cord?

Line Cord Maintenance

1. Is the insulation covering of the line cord cracked? Are the wires exposed? Is the cord positioned so it can be stepped on often? Easily?

2. Is the plug molded onto the line cord? If yes, does the connection appear to be loose? If the plug is nonpolarized, shouldn't a polarized type be substituted for it?

3. Is the line cord long enough that no strain is put on it when it is connected?

4. Does the line cord make good contact with the device to which it delivers electrical power?

5. Is the line cord capable of carrying the current required by its connected electrical device?

6. Does the line cord feel warm or hot to the touch after being in use for a short time?

7. Is the line cord placed so no one can trip over it?

8. Is the line cord one that has been spliced to extend its length?

Extension Cord Maintenance

1. Is the cord designed for the amount of current it is expected to handle?

2. Is the cord rolled up and stored out of the way when not in use?

3. Does the insulation of the cord have any cracks, rips or tears anywhere along its length?

4. Is the plug of the cord in good condition?

5. Is the receptacle of the cord in good condition?

6. Does the cord get hot during the time it is used? Does its receptacle also get hot?

7. Do electrical components work properly when used with the cord?

8. Is the cord too long? Too short?

9. Is the cord subject to abuse when it is used? Do people step on the cord? Is the cord left outdoors overnight?

10. Is the cord an old one that uses a nonpolarized plug?

11. Has the cord been extended by adding a length of cable to it?

12. Is the cord an old one that does not have a ground wire?

13. Is the cord so old that its insulation has lost its flexibility?

14. Are cords of various lengths available throughout the home?

15. Has the insulation of the cord ever been burned by a soldering iron or accidentally cut by a tool?

16. Do you have a drop cord, an extension cord equipped with a light and receptacle at one end?

17. Is the drop cord light in good working condition?

18. If the extension cord is seldom used, do you rewind it on its reel from time to time?

19. Is the line cord placed where it is readily accessible?

20. If the line cord is equipped with a light, does its switch work properly?

21. Are you using line cords as a means of avoiding rewiring or adding new electrical branches?

22. Do you connect the line cord to a branch that is near the limit of its current-carrying ability?

23. Do you ever use the line cord near machinery that could cut and short the cord?

24. Is the line cord ever used in a hostile environment?

GENERAL MAINTENANCE FOR ALL MOTORS

1. Motors are designed to work with specific loads. For vacuum cleaners, empty the dirt bag regularly or replace it. When the bag is full, or almost so, the motor must work harder. Washing machines and dryers operate with specific loads, measured in pounds. Use a weight scale to check.

2. Motors have frames equipped with ventilation slits. Vacuum the slits or use long tweezers to remove dirt. Dirt can interfere with turning of the rotor.

3. Some motors are equipped with oilless bearings which never need lubrication. If not, oil the bearings with one or two drops of oil. Do not use too much oil, as this will result in the buildup of grime and will attract and hold lint. Use only the oil recommended by the manufacturer.

4. If the motor produces a laboring sound, it may be overloaded. Check with a much lighter load to determine if the sound reduces or disappears.

5. If the operating strength of the motor seems to vary, check the belts and pulleys. These may be loose. Examine the belts for cracks and wear.

6. Motor brushes may be sparking excessively. A small amount of sparking is normal. It may be necessary to readjust the brushes. If too close to the commutator or slip rings, the brush holders may be scratching against them. Replace the brushes if they are excessively worn.

7. Poor motor operation can be caused by excessive dirt on the commutator segments or on the slip rings. Clean with fine sandpaper or a small piece of canvas. Do not use emery cloth.

TROUBLESHOOTING FLUORESCENT LIGHTS

Lamp Does Not Light

Make sure the lamp is connected to a live power source. The lamp may not be seated properly in its end connectors, or it may be making poor contact. Try rotating the lamp while the power is on. If the lamp shows blackened areas at its ends, then the lamp is defective. Substitute a known good lamp. Some fluorescents will not light if the room temperature is too low. On older types of fluorescents the starter or ballast may be burned out.

Lamp Lights But Is Dim

The lamp may be dirty. Remove and clean with a damp cloth. Wipe thoroughly and make sure the lamp is dry before reinserting it in the fixture. The lamp may be defective. Try replacing it with a known good lamp.

Lamp Blows Fuse or Opens Circuit Breaker

The branch to which the lamp is connected may be overloaded.

Humming Sound From Fluorescent

This problem generally exists with older fluorescent fixtures using ballast. The wrong ballast may be used, may be incorrectly installed, or may be old and noisy. Ballast replacement is needed.

Lamp Flickers

This condition can exist with new lamps. Keep using the fixture and let it remain turned on for several hours. The end connectors of the lamp may not be well seated. If tube has been in use for a long time, it may be nearing the end of its life. This problem sometimes exists in winter in unheated rooms. The lamp socket may be loose.

Fluorescent Light May Have Short Life

This situation can be caused by excessive use of the on-off switch.

Fluorescent fixtures are capable of causing television interference (TVI) or radio interference (RFI). This is because fluorescents behave like miniature noise transmitters.

A cure may not always be possible, since the fluorescent lights transmit electrical noise in two ways: Through the power line and through the air, as in the case of a broadcast station. To eliminate RFI and TVI, move the radio receiver and the television receiver as far from the fluorescent fixture as possible. Also try connecting the radio and TV sets to different receptacles. Noise filters are sold that are supposed to reduce electrical noise. They are plugged into the receptacle, then the radio or TV sets are connected to the filter. There is no assurance that such filters will work. FM reception is much less susceptible to RFI that is air transmitted. The interference will vary from one end of the tuning range of the FM set to the other, so it may be possible to get noise-free reception at one end of the FM dial.

TROUBLESHOOTING INCANDESCENT LIGHTS

Light Does Not Turn On

If the bulb is brand new, do not assume it cannot be defective. Try another bulb known to be in working order. Check the receptacle to determine if it is supplying power. Remove fuse or open circuit breaker, open switch or remove plug from its receptacle. Examine the interior of light socket. Sometimes the center contact of the light socket is pushed down. Lift it a short distance. If the nonfunctioning light is part of a group on the same fixture and the other lights are working, the fault is in the socket or in the bulb. The switch controlling the light may be defective.

Light Works But Seems Dim

The bulb may be dirty. Remove, wash, and then dry throughly before replacing. The bulb may be a lower wattage type. It is sometimes difficult to read the wattage rating on the bulb. Replace the bulb with a new one whose wattage is printed on its container. Are frosted bulbs being substituted for clear types? Bulbs using color have less light output per watt.

Light Bulbs Seem to Have a Short Life

AC line voltage is nominally 120 volts. The voltage in your location may be higher. Check with your local power utility.

Light Is Intermittent

The bulb is loosely seated in its socket.

HOW TO REPAIR TABLE AND FLOOR LAMPS

Pull-chain type lamps are constructed using a socket, as shown in Figure 8-1. If the problem is a light that is intermittent or one that does not work at all, even when using a new bulb, remove the connecting lamp cord from its receptacle. Sometimes

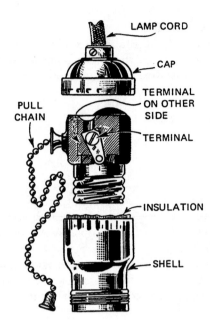

Figure 8-1 Pull-chain socket assembly of lamps.

the center contact of the socket has been pushed too far down. Reach into the socket and pull this contact up about 1/8 inch. After making this adjustment, replace the bulb and insert the male plug into the receptacle.

A more common problem is degeneration of the lamp cord, its plug, or a combination of the two. If the cord is an older type made of rubber, its age may have caused its insulation to crack. To remove the existing lamp cord, separate the socket into its component parts. Adjacent to the pull chain are two terminals to which the exposed ends of the new lamp cord can be connected. The cord will probably consist of stranded wire. No soldering will be necessary. Strip about 1/2-inch of insulation from the two wires and try not to cut any of the strands. Twirl the strands, always in the same direction, to supply the equivalent of a single wire. Wrap this wire around the terminal screw in a clockwise direction and tighten the screw. If any of the strands extend beyond the screw, cut them away, using diagonal cutters. Repeat this procedure with the other screw.

Make sure that the insulation on the inside of the shell covers the wiring just completed. The top edge of the shell is ridged and so is the bottom edge of the socket cap. These two parts will snap together when they are pushed against each other.

Not all pull-chain sockets are made in the same way. Any differences will be mechanical, but the basic electrical connections will be the same.

There are two types of plugs used with lamp cord: molded and separate. The molded eliminates the need for connecting the plug to the cord. However, a separate plug may be required if a combination cord and plug are not available. When they are available, it may be necessary to cut the cord to a suitable length.

THE PROBLEM OF COLOR CODING

In the electrical industry white is the standard color for a neutral wire, while black and red are used for the hot leads. The advantage is that the funtion of each of the wires is known at once, without the need for testing.

Figure 8–2a shows a switch and a lamp used in an AC circuit. The switch is in series with the black lead, identified by the letter B. The lamp, though, is placed across the AC line, and one of its terminals is connected to the hot lead; the other terminal to the white.

Part (b) shows an alternative circuit arrangement and is one in which the two parts, the lamp and the switch, have been transposed. In either circuit, though, when the switch is closed, the lamp is shunted across the AC line.

While both circuits in Figure 8–2 will work, that in part (a) is preferable. If, at some future date, the switch needs to be pulled out of its receptacle and examined, the wiring of (a) will not be questionable, for it follows standard wiring procedure. A single-pole, single-throw switch has only two terminals, with a black lead connected to one and also to the other. This emphasizes that the black lead has been cut open, and the switch has been inserted in series with that lead.

Following (b), however, examination of the switch will show a black wire connected to one terminal and a white wire to the other. Since this can cause confusion, some electricians, when using circuit (b), paint the white lead going to the switch a black color to emphasize that it is actually a part of the hot lead. Using circuit (a) eliminates this cautionary procedure.

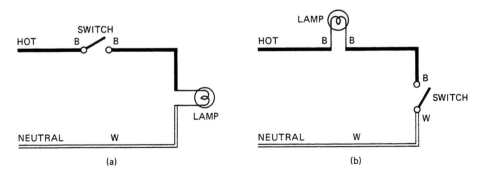

Figure 8-2   Two possible circuit arrangements for switches. B (black); W (white).

SERIES AND SHUNT CONNECTIONS

Electrical wiring consists of two types of connections: series and shunt. With a series connection the black (or other hot lead) is opened and a component is inserted in series with it. All switches are wired in series with a hot lead. The switch becomes part of this lead and is used to permit current to flow or not to flow.

A shunt connection, unlike the series, does not involve breaking into a line. Instead the component is connected across two lines, from the black (hot) lead to the white (neutral) lead. All components that require current—lamps, radio or TV sets, washing machines, motors—are wired in shunt. Considering a single branch, all components that are plugged into all receptacles serviced by that branch are automatically connected in shunt with the white and black leads and also with each other.

Shunt Rules

1. Never connect a fuse in shunt with the power lines.
2. Never connect a switch in shunt with the power lines.
3. Never connect a wire, or any other current-carrying bit of scrap metal in shunt with the power lines. Motors and transformers are an exception to this rule, even though they consist of conductors. Motors and transformers are designed to work in parallel with the power line; a length of wire or a bit of scrap metal are not.
4. When an appliance is plugged into a receptacle, it is automatically put in shunt with the power line.
5. Since all applicances are in shunt with the power line, they are also in shunt with each other.
6. The larger the number of active appliances plugged into receptacles, the greater the total load and the greater the total amount of current taken from the line.
7. The maximum load on a power line is a short circuit and consists of the joining the black and white leads. Whatever the cause of the short circuit, the result is opening the fuse or circuit breaker.
8. Never connect an unknown device or component into a receptacle just to see what will happen.
9. Never try to force a two-prong polarized plug into a two-terminal nonpolarized receptacle.

Series Rules

1. Never wire appliances in series unless they are specifically intended to be connected in this way. Some Christmas tree lights come into this category.

2. Certain components must always be wired in series with the black lead of a branch. These include fuses, circuit breakers and switches.

3. If two components, such as 120-volt light bulbs, are connected in series, the voltage across each will be 60 volts. Under these circumstances the bulbs will quite possibly not light.

4. When two components are wired in series and one of them is removed from its receptacle, the other will not work. Removal of the component is equivalent to the action of a switch.

How to Locate Blown Fuses or Open Circuit Breakers

Whether you have a circuit breaker or fuse-type service box, the first step in checking for trouble is to look. Don't touch, just look.

If you have a fuse box, examine each fuse carefully using a flashlight, even though the basement or closet is lighted. The flashlight will supply concentrated, instead of dispersed, lighting.

The blown fuse may have a dark smear across its top (on the inside). On some fuses the conducting element is held in position near the top. If it is no longer there, the fuse is open.

If you have a circuit breaker service box, the work is easier. Examine the trip handles of each switch. If you cannot decide from a visual inspection whether any of the breakers have tripped, run a finger along the handles. They should all be in about the same position. If one of them happens to be out of line, put it back into its on position. Pull its handle back to off and then return it to on.

Fuse or Breaker Problems

Theoretically, you should never have to replace a fuse or reset a circuit breaker. But if a particular fuse or breaker keeps opening, you are being given a warning signal that something is wrong. Don't ignore it!

Remove the fuse and examine it. Is it the right size for the branch? Does that branch call for a 20-ampere fuse which has been replaced with a 10-ampere unit just because it happened to be handy?

Examine the window of the blown fuse. Is it severely discolored? Is there a black smudge across its face? If yes, then there is a short circuit somewhere in the branch it protects.

The next step is to locate the cause of the short. It could be in the branch feeding a receptacle, in a receptacle or in an appliance. Remove all appliances connected to the receptacle (or receptacles) being fed by that power branch. Replace the fuse for that branch, being sure to use one having the correct rating. If the fuse does not blow again, then the branch wiring isn't shorted. Replace each appliance in the receptacle it normally uses. Do not use more than one appliance at a time. Check the fuse each time a replacement is made.

If one of the appliances blows the fuse, then it has a short. If none of the appliances affects the fuse, then it is possible the branch line is being overloaded.

If the fuse blows with no appliance inserted in any of the receptacles in the branch, there is a short in the power line, or, more likely, a short in one of the receptacles. Open each box containing a receptacle and examine the wiring carefully. Look for evidence of a short, such as a black smudge or charred insulation.

TESTING WITH YOUR SENSES

Many of the trouble symptoms supplied by house wiring and components are subtle, but they can often be detected by your being electrically alert. Excessive sparking at the brushes of a motor, a plug that feels hot, a television picture that has pulled in from the top and sides, lights that dim when a washing machine is turned on, a burning odor from a receptacle, an extension cord that feels unusually warm, a plug that needs to be wiggled back and forth before an appliance will work, an excessive number of receptacle taps, repeated openings of fuses or circuit breakers, a number of appliances that operate intermittently—all of these and more mean that the electrical system is trying to call attention to itself.

Because some of these symptoms aren't ongoing, it is often easy to ignore them, hoping they will go away. That they will not do. In a best case situation they will cause loss of electrical power in one or more branches. In a worst case situation they will result in a fire.

To service the problem, consider that the wiring system consists of a number of branches and that it is unlikely that all the branches are affected. The first step is to localize the branch, to determine which branch is the troublemaker. The next step is to find which component in that branch is at fault. In some instances the only fault is that the branch is overloaded. If not, then check all receptacles and switches in the branch being investigated.

BRANCH LINE SERVICING

The electrical wiring in a home consists of a number of branch lines, all connected to fuses or circuit breakers in the service box. The electrical power supplied by a local utility enters this box, is divided into branches, and is then routed by cables throughout the house. Each of these branches may be regarded as independent, since a fault in one will not often affect any other.

The ideal way to maintain and update the wiring in a home is to avoid waiting until some electrical fault becomes evident. To wait would mean suffering the inconvenience of an electrical failure. The best method is to make a sketch of the house wiring plan.

A branch consists of its fuse or circuit breaker, the wiring of that branch, the junction boxes, and the other electric boxes containing switches and receptacles. In some instances the failure of a receptacle to supply power may have nothing to do with that particular receptacle but may be caused by some other receptacle in the branch. Having a plan of the branch and the location of all the junction boxes and receptacles in that branch can simplify fault finding.

If, for example, power is delivered to one receptacle, there is a continuation of the power line to a second. But if the power line from receptacle 1 to receptacle 2 becomes disconnected in receptacle 1, a puzzling situation arises. Receptacle 2 has no power, while receptacle 1 remains active. As a result appliances plugged into receptacle 2 do not operate. The usual approach is to test the appliances for defects, to remove receptacle 2 from its electric box, to remove the wires from receptacle 2, and to test them for voltage. This is a substantial amount of unnecessary work, since it has not resulted in a location of the fault.

A sketch of the branch could have indicated a fault possibility. A neon tester would have revealed that receptacle 1 was live but that there was no voltage supplied to receptacle 2. The drawing of the branch would have indicated immediately that receptacle 2 was supposed to receive its power from receptacle 1. Therefore, even though receptacle 1 was live, it was the location of the fault. Quite possibly one of the exit wires from that receptacle, a push-in type, had become disconnected.

RECEPTACLE PROBLEMS

One of the advantages of a receptacle is that, unlike a switch, it has no mechanically moving parts. However, it can produce problems of its own.

Faulty Mounting

The receptacle has a pair of machine screws in front, mounted in horizontally positioned slots. This means there is some latitude in the way the receptacle can be inserted into its electric box. If the receptacle has side mounting screws, the wires connected to the screws can accidentally touch the sides of the box. If the box is made of metal, this can result in a short. Usually the distance between the sides of the box and the screws of the receptacle is quite small, so the receptacle must be positioned carefully.

When attaching wires to the screws of the receptacle, do not allow any wire to extend beyond the screw head. Doing so serves no useful purpose and can be hazardous. Use solid wire only, never stranded wire, for making connections to a receptacle. No matter how carefully stranded wire is twirled to form the equivalent of a solid conductor, it is always possible for a single strand to escape. All branch wiring should be solid conductor.

When connecting wires to a receptacle, allow some extra length to permit removing the receptacle, should that become necessary. The extra length also makes it easier to wire connections to the receptacle when it is first installed. Since the wire is well insulated there is little chance that it will short against the box.

Most receptacles are duplex types. The screws for the neutral wire are silver colored and are connected to each other by a short strip of metal. This means that the neutral wire can be continuous, even if it is cut and connected to either of the silver screws. The black or hot lead is connected to a brass screw. Do not transpose the wires, and be sure to attach them to the proper screws.

A plug must make firm contact when inserted into the receptacle. If the plug is loose, it can result in no operation or poor functioning of the connected appliance. To determine whether it is the plug or the receptacle which is at fault, try another plug. If it makes a firm fit and the appliance still does not work, the receptacle is defective and should be replaced. Try the original test plug in several receptacles to make sure that its prongs are not at fault.

HOW TO CHECK FOR A LIVE RECEPTACLE

Before removing a receptacle, make sure it is no longer capable of delivering electrical power by removing its fuse or inactivating its circuit breaker. That is a first step, but it is not enough. You may have removed the wrong fuse or worked on the wrong breaker.

There are a number of ways of checking a receptacle for power availability and all of them are good.

An easy way is to use an extension cord equipped with a test lamp. First check the cord and its light by plugging it into a known active outlet. Turn the lamp switch to its on position and keep it that way. Now plug it into both bosses of the receptacle, assuming it is a duplex type. A *boss* is the extrusion of a receptacle and is the part into which a plug is inserted. If the light does not turn on, the receptacle is inactive.

In the absence of a light-equipped extension cord, use a portable light, following the same procedure. Be sure to turn the light on and then check each boss in turn. For each of these tests the light of the test lamp should remain turned on.

These tests indicate whether the receptacle is electrically active or not. Another method is to use a neon-type tester. Plug the tester into both access slots. The lamp should remain turned on. Keep a test lead in one of the slots, and touch the center wall plate screw with the other. Then transfer the test lead to the other slot, while keeping the test lead on the center screw. In one of these tests the neon lamp should remain on.

Another technique is to use a volt-ohm-milliammeter. Set the function switch of the instrument to AC volts, and set the range selector so that 120-volts comes somewhere in the center of that range. Insert the test prods into both slots to see if any voltage registers on the scale of the instrument. If no voltage is indicated, the receptacle is inactive. Repeat the test with one test prod on the center screw of the wall plate and the other test prod inserted into each receptacle slot in turn. Again, voltage should be indicated in one test; no voltage in the other.

HOW TO TEST THE CENTER SCREW OF A RECEPTACLE PLATE
FOR GROUND

The center screw holding the wall plate of a metal receptacle in position should be grounded. If it is, then this is an indication that the electric box is also grounded.

This requires a live test. If the fuse has been removed or the circuit breaker inactivated, restore them to their active condition. Using a neon bulb tester, insert one test prod into the slot of a receptacle and then touch the other lead to the center screw. Keep the test prod on the center screw and insert the test prod into the other receptacle slot. With one of these tests, the neon test lamp should light. If it does, the center screw is grounded. If the test lamp lights when the test prods are inserted in both slots, but not when one of the prods touches the screw, then the screw is not grounded. Assuming the electric box is metal and not plastic, this is an indication that the ground lead of the incoming power cable has not been connected to the box.

If these tests show that the receptacle is not grounded, remove the fuse or open the circuit breaker at the service box. Unscrew the center holding screw of the faceplate and then the two screws holding the receptacle in the electric box. Pull the receptacle out and locate the ground wire of the incoming cable. This should be a bare wire or, if insulated, color coded green. Make sure this wire is connected to the ground clip or to the ground screw of the receptacle. This connection may have been omitted accidentally.

HOW TO TEST A SWITCH WITH A NEON LAMP

To test a switch with a neon lamp, connect its leads across the terminals holding the white and black wires. If the tester is across the input terminals (across the cable bringing power in), the neon lamp should glow regardless of the setting of the switch.

Now transfer the test lead to the screw holding the hot output wire. This is the wire leading to the appliance. The test lead connected to the white (neutral) wire need not be moved. When the switch is turned on, the neon lamp should glow if the switch is in good working order. When the switch is turned off, the glow of the neon lamp should disappear.

Audible Switch Testing

Some switches are silent types and produce no sound when operated. A switch of this type, for example, might be desirable in a room or area used by a sleeping infant. Other switches make a definite clicking sound when operated. If it does not do so, then it is likely the switch is defective, particularly if the light or other device it controls doesn't work.

To test, remove the faceplate and make a voltage test with a neon lamp tester. If the switch is recessed and the wired connections aren't accessible, turn off its circuit breaker or remove its fuse. Unscrew the two holding screws of the switch and pull it out of its receptacle. Restore power at the service box by reinserting the fuse or resetting the circuit breaker. Test across the black and white input wires and then across the black and white output wires. If there is voltage across the input but none across the output, regardless of the switch setting, the switch is defective.

Replace the switch with a new unit, making certain to reconnect the wires as they were positioned originally.

TROUBLESHOOTING DOORBELLS, BUZZERS OR CHIMES

System Does Not Work

The most common cause of sound failure is the front door pushbutton. If the system has a rear door pushbutton and it works, then the front door pushbutton is at fault. To check, pull out the pushbutton, but do not disconnect the wiring, and short the terminals to each other. This should operate the system. If it does, replace the pushbutton with a new unit. The transformer may be faulty, although this seldom happens. Make sure the primary of the transformer is connected to the power source and that it is receiving line voltage. Check with a neon bulb tester.

If primary is receiving voltage, check secondary of the transformer with a volt-ohm-milliammeter (VOM) set to read low volts AC. Bells and buzzers require 6 to 10 volts; chimes are rated at 15 to 20 volts. If there is no voltage across secondary or if it is very low, the transformer is defective. If there is voltage and it is the correct amount, there may be an open in the wiring leading to the pushbuttons.

In some instances the inoperative fault may be due to the bell, buzzer or chime. In the cases of a bell or buzzer, try substituting a new unit. Chimes can be solenoid operated. The solenoid rod must be able to move back and forth. Watch the solenoid as the pushbutton is being operated. If the solenoid rod doesn't move the chime will not function. The repair may require a new solenoid from the distributor or manufacturer.

HOW TO TEST A TRANSFORMER

There are several ways of testing a transformer, but the best is one that does not require disconnecting the unit from its voltage source. It does need the assistance of a volt-ohm-milliammeter (VOM). Set the instrument's function control to read

AC volts and adjust the range selector for a mid-scale deflection of 120 volts. Put the test leads across the primary wires. A reading of approximately 120 volts indicates that the transformer is receiving its required input voltage, that the branch power line connected to the primary winding of the transformer is working properly, and that the fuse (or circuit breaker) is in good working order. This can be considered a satisfactory number of results from a single test.

Keep the function selector of the VOM set for AC volts, but lower the range selector for a maximum reading of about 50 volts. Connect the test leads of the VOM across the secondary terminals of the transformer. A voltage reading between about 6 to 20 volts indicates that the transformer secondary is supplying voltage for a bell, buzzer or chimes and that the transformer is working.

When using the VOM, keep your fingers away from the metal portion of the test leads to avoid receiving a shock.

A neon bulb tester can be used, but it can only indicate the voltage across the primary winding. It will not respond to the secondary voltage.

If the transformer has an odor of burning insulation or if it feels hot to the touch, turn the power off, remove the transformer, and substitute a new unit.

TROUBLESHOOTING THE OUTDOOR ELECTRICAL SYSTEM

The outdoor electrical system consists of two branches: one of these is low voltage, transformer operated, while the other is directly connected to the 120-volt line. These branches are independent, thus simplifying repairs.

The Low-Voltage System

Problems in the low-voltage system usually involve a single bulb which does not light. This is solved by replacing a bulb with one that is known to be good. If all the bulbs but one light, then the transformer, its connecting AC line, and its fuse or circuit breaker are in good working order, and the problem is generally caused by the connecting wire's separating from the base of the spike that supports the unlit bulb. Even though the voltage is low, turn off the circuit breaker or remove the fuse that protects the low wire line. Dig around the spike and locate the connecting wire. Check the wire and make sure it is making contact with the wires in the spike.

If none of the lights work, the problem involves the transformer, the connecting branch line back to the fuse or circuit breaker box, or possibly the line from the secondary of the transformer to the cable leading to and connecting to the bulbs.

If the transformer has a burning odor or feels hot to the touch, it may need to be replaced. The transformer may have an internal short, or there may be a short anywhere along the line from the secondary of the transformer to the bulbs.

Check the connections to the primary and secondary windings of the transformer, and make sure these are secure. Check the fuse or circuit breaker.

Checking the 120-volt Outdoor Branch

Side door lights, post lantern lights, flood lights, and infrared security lights are all 120-volt operated and, unlike the low-voltage lighting system, are not transformer operated. All of these devices lead to a junction box inside the home and to separate indoor switches. There may also be one or more outdoor receptacles connected to the outdoor cable.

The most common fault in the lighting system is due to one or more faulty bulbs. If replacement using a bulb known to be in good working order does not restore lighting, check the fuse or circuit breaker. If these are in working order, check for the presence or absence of voltage at the socket holding the nonfunctioning light. You can use a neon light tester or a VOM for this, but be careful. If voltage is not present, move back to the indoor receptacle box, and remove its cover plate. If the wires are covered with wire nuts, shut off power by disabling the fuse or circuit breaker. Remove the wire nuts and expose the wires. Restore power and then check for voltage. A wire may have come loose.

If voltage is still missing, trace the wiring back toward the service box. It may be that the voltage is being supplied from a junction box. Remove the cover plate and then run a test following the same procedure as with the receptacle box that has just been checked.

Checking an Outdoor Receptacle

If all the 120-volt outdoor lighting works but an outdoor receptacle does not supply voltage, check the connections to the receptacle, following the same procedure as that described for checking the indoor receptacle.

Working with the 120-volt line outdoors is riskier than checking the same voltage inside, since the chances of outside moisture are so much greater. Do this work only when conditions are dry. Make sure your shoes and socks are also dry. And be certain the surface you are standing on is dry.

Index